T0298011

Combinatorial
Extremization

Mathematical Olympiad Series

ISSN: 1793-8570

Series Editors: Lee Peng Yee *(Nanyang Technological University, Singapore)*
Xiong Bin *(East China Normal University, China)*

Published

Vol. 4   Combinatorial Problems in Mathematical Competitions
by Yao Zhang (Hunan Normal University, P. R. China)

Vol. 5   Selected Problems of the Vietnamese Olympiad (1962–2009)
by Le Hai Chau (Ministry of Education and Training, Vietnam)
& Le Hai Khoi (Nanyang Technology University, Singapore)

Vol. 6   Lecture Notes on Mathematical Olympiad Courses:
For Junior Section (In 2 Volumes)
by Xu Jiagu

Vol. 7   A Second Step to Mathematical Olympiad Problems
by Derek Holton (University of Otago, New Zealand &
University of Melbourne, Australia)

Vol. 8   Lecture Notes on Mathematical Olympiad Courses:
For Senior Section (In 2 Volumes)
by Xu Jiagu

Vol. 9   Mathemaitcal Olympiad in China (2009–2010)
edited by Bin Xiong (East China Normal University, China) &
Peng Yee Lee (Nanyang Technological University, Singapore)

Vol. 11   Methods and Techniques for Proving Inequalities
by Yong Su (Stanford University, USA) &
Bin Xiong (East China Normal University, China)

Vol. 12   Geometric Inequalities
by Gangsong Leng (Shanghai University, China)
translated by: Yongming Liu (East China Normal University, China)

Vol. 13   Combinatorial Extremization
by Yuefeng Feng (Shenzhen Senior High School, China)

The complete list of the published volumes in the series can be found at
http://www.worldscientific.com/series/mos

Yuefeng Feng

Shenzhen Senior High School, China

Vol. 13 | Mathematical Olympiad Series

Combinatorial Extremization

East China Normal University Press

World Scientific

Published by

East China Normal University Press
3663 North Zhongshan Road
Shanghai 200062
China

and

World Scientific Publishing Co. Pte. Ltd.
5 Toh Tuck Link, Singapore 596224
USA office: 27 Warren Street, Suite 401-402, Hackensack, NJ 07601
UK office: 57 Shelton Street, Covent Garden, London WC2H 9HE

British Library Cataloguing-in-Publication Data
A catalogue record for this book is available from the British Library.

Mathematical Olympiad Series — Vol. 13
COMBINATORIAL EXTREMIZATION

Copyright © 2016 by East China Normal University Press and
World Scientific Publishing Co. Pte. Ltd.

ISBN 978-981-4730-02-0
ISBN 978-981-4723-16-9 (pbk)

Printed in Singapore.

Introduction

This is a book devoted to methods of discrete extremization. The reader may be familiar with some of the methods presented here, but not with the others. Therefore, there are two choices in reading this book.

One choice is to read about the familiar methods first, so that it is relatively easy to transit to other methods. The other choice is to begin with the less familiar ones, since this may be more effective.

It is more important for the reader to focus on the general idea and thoughts, instead of being overwhelmed by the details of each problem.

The reader is advised to fully master the basic steps of the methods and the use of each step, and to improve understanding through examples. Next, based on the understanding of the methods, the reader can try to solve sample problems in the book before reading further, and compare one's own methods with the solutions provided in this book to examine one's understanding of the methods. Finally, the reader is encouraged to connect the techniques of this book with problems encountered earlier, and to try to solve those problems using the methods presented in this book, in order to firmly understand the methods and use them flexibly in practice.

Preface

Extremization problems have always played an important role in mathematical competitions, while combinatorial extremization problems are always among the hardest of all extremization problems.

The meaning of combinatorial extremization is: suppose there is a function with arguments being natural numbers or integers, or a problem involving discrete variables such as sets, subsets and elements. It is then required to find the extremum of the function or some corresponding expression, under certain specific conditions (not necessarily in the form of equalities or inequalities, but probably certain *"discrete"* properties of the arguments).

A combinatorial extremization problem usually includes two aspects: proof and construction. *"Proof"* is to justify that a certain variable satisfies some inequality or a certain object has a certain property. *"Construction"* is to construct an object satisfying the requirements given by the problem, or a counterexample to disprove the statement of the proposition. These two aspects are normally quite different regarding both the perspective of thinking and the ways of problem solving, and they both require flexible thinking, rich imagination and creative idea. Therefore they are often emphasized in mathematical competitions.

Combinatorial extremization problems can be divided into two types: the *"sum-product type"* and the *"parameter type"*.

The *"sum-product type"* extremization problem refers to finding the extremum of quantities which can be expressed by a function with a *"sum"* or *"product"* structure. In the first five chapters of this book, we introduce five common methods in solving these kinds of problems.

The *"parameter type"* extremization problem refers to extremizing some parameter in the problem. The main feature of this type of problems is that the quantity to be extremized has no explicit expressions. In the last eight chapters of this book, we introduce eight common methods in doing these kinds of problems.

The parameter type extremization problems can further be divided into two categories: *"existence"* extremization problems and *"forall"* extremization problems.

By an *"existence"* extremization problem, we mean a problem that requires finding the extremum of some parameter k so that some object with property p *"exists"*. Here the *"proof"* part is to prove that, if $k > k_0$ (or $k < k_0$), any object satisfying the assumption will not have property p, deducing that $k \leqslant k_0$ (or $k \geqslant k_0$).

The *"construction"* part is to construct an object with property p with $k = k_0$. Roughly speaking, to solve an *"existence"* extremization problem, one needs to prove certain inequalities in order to obtain a range for the parameter, and construct an example in order to demonstrate that equality can hold.

On the other hand, a *"forall"* extremization problem refers to a problem where one is required to find the extremum of a parameter k, such that some property p holds *"forall"* objects. The *"construction"* part is then to prove the statement: "if $k > k_0$ (or $k < k_0$), there exists an object satisfying the condition but does not have the property p." Therefore k satisfies inequality $k \leqslant k_0$ (or $k \geqslant k_0$). The *"proof"* part is to prove that, when $k = k_0$, indeed all objects have property p. In a word, *"construction"* is used to obtain an inequality, and *"proof"* is used to show that equality can hold. Please note of the difference of the methodologies in these two kinds of problems.

Although different chapters of this book are related to each other to a certain extent, they are also relatively independent. It is not necessary for the reader to follow the order of this book; one may choose the parts that one is interested in. The reader can also leave the difficult problems aside and go back to these problems while one has a comprehensive understanding of the methods discussed. Each chapter has some practices, which usually can be solved using the methods introduced in this chapter, and the answers are also attached at the end of this book for the reader's reference. The reader is encouraged not to be confined to the solution, and to try to find more creative solutions.

The book is suitable for high school students, teachers and students in the Department of Mathematics of normal colleges, and anyone interested in Mathematical Olympiads.

Limited by the knowledge of the author, this book will inevitably contain errors. Readers are welcome to point them out.

Contents

One of the most significant characteristics of combinatorial extremization problems is that the constraints or the expressions of the target function are rather complicated. The so-called *inequality control* method is to enlarge or shrink the constraints or the target function, in such a way that the relation between the assumptions and the goal becomes more apparent. By enlarging or shrinking, the problem will become close to a standard form, where one needs to find the extremum of $u = g(x, y)$ given $f(x, y) = 0$. In this way, the combinatorial extremization problem is transformed into an analytic extremization problem.

There are usually two ways to use this method. One is to enlarge or shrink the constraints, making clear the subtle or hidden constraints; the other one is to enlarge or shrink the target function, simplifying the complicate expressions of the function.

Example 1. Assume there are m different positive even integers and n different positive odd integers that add up to 1987. Find the maximum of $3m + 4n$. (2nd CMO)

Analysis and Solution. The difficulty of this question lies in the fact that the constraints are complicated. They can be simplified via inequalities, and then be enlarged or shrunk to meet the target function.

Assume that the m given positive even integers are $a_1, a_2, \ldots, a_m$ and the n positive odd integers are $b_1, b_2, \ldots, b_n$, then

$$(a_1 + a_2 + \cdots + a_m) + (b_1 + b_2 + \cdots + b_n) = 1987. \qquad \textcircled{1}$$

Notice that the target function is a function of m, n, and in the constraints m, n only act as subscripts. Thus the constraints in a_1, $a_2, \ldots, a_m$ and $b_1, b_2, \ldots, b_n$ in ① should be transformed into constraints in m, n.

Since $a_1, a_2, \ldots, a_m$ and $b_1, b_2, \ldots, b_n$ are mutually different positive even integers and positive odd integers, we have

$$
\begin{aligned}
1987 &= (a_1 + a_2 + \cdots + a_m) + (b_1 + b_2 + \cdots + b_n) \\
&\geqslant (2 + 4 + 6 + \cdots + 2m) + (1 + 3 + \cdots + 2n - 1) \\
&= m^2 + n^2 + m.
\end{aligned}
\tag{②}
$$

Notice that our goal is to prove $3m + 4n \leqslant A$ for some A, which is akin to the form of Cauchy-Schwartz inequality. Thus the right-hand side of equation ② should be transformed into a *"sum of square"*. We have

$$
1987 + \frac{1}{4} \geqslant \left(m + \frac{1}{2} \right)^2 + n^2
$$

$$
\left(1987 + \frac{1}{4} \right)(3^2 + 4^2) \geqslant (3^2 + 4^2)\left(m + \frac{1}{2} \right)^2 + n^2
$$

$$
\geqslant \left(3\left(m + \frac{1}{2} \right) + 4n \right)^2.
$$

Therefore

$$
3m + \frac{3}{2} + 4n \leqslant 5\sqrt{1987 + \frac{1}{4}}.
$$

This implies that

$$
3m + 4n \leqslant \left[5\sqrt{1987 + \frac{1}{4}} - \frac{3}{2} \right] = 221.
$$

Now we will construct an example where equality holds. First we find the pair of (m, n) such that $3m + 4n = 221$.

This Diophantine equation has many solutions. In order to find the pair (m, n) which satisfies equation ②, the corresponding even and odd integers should be as small as possible, which requires that m and n are sufficiently close. Enumeration then gives $m = 27$ and $n = 35$, for which $3m + 4n = 221$ and $m^2 + n^2 + m = 1981 < 1987$, which satisfies equation ②.

Choose the smallest 27 positive even integers $a_1 = 2$, $a_2 = 4, \ldots$, $a_{27} = 54$, and 35 smallest positive odd integers $b_1 = 1$, $b_2 = 3, \ldots$, $b_{34} = 67$, $b_{35} = 69$, then

$$(a_1 + a_2 + \cdots + a_{27}) + (b_1 + b_2 + \cdots + b_{35}) = 1987 - 6.$$

Then modify b_{35} to: $69 + 6 = 75$, we obtain

$$(a_1 + a_2 + \cdots + a_{27}) + (b_1 + b_2 + \cdots + b_{35}) = 1987.$$

In summary, the maximum of $3m + 4n$ is 221.

Note: the key to solving this problem is to transform ① into ②, while the use of Cauchy-Schwartz inequality is not crucial. In fact, after equation ② is obtained, we can also use a trigonometric substitution to find the extremum of $3m + 4n$:

From ②, set

$$r = \sqrt{1987 + \frac{1}{4}}, \quad m = -\frac{1}{2} + r\cos\theta, \quad n = r\sin\theta,$$

then

$$3m + 4n = \left(-\frac{1}{2} + r\cos\theta\right)^2 + (r\sin\theta)^2 + \left(-\frac{1}{2} + r\cos\theta\right)$$

$$= 3r\cos\theta + 4r\sin\theta - \frac{3}{2} = 5r\sin(\theta + t) - \frac{3}{2}$$

$$\leqslant 5r - \frac{1}{2} \leqslant 5\sqrt{1987 + \frac{1}{4}} - \frac{3}{2} \text{ (similarly hereinafter)}.$$

Example 2. Let $x_1, x_2, \ldots, x_n \in \mathbf{R}^*$, $\sum_{i=1}^{n} \frac{1}{x_i} = A$ be fixed. Given a positive integer k, find the maximum of $\sum \frac{1}{x_{i_1} + x_{i_2} + \cdots + x_{i_k}}$ where the summation is taken over all k-element subsets $\{i_1, i_2, \ldots, i_k\}$ of $\{1, 2, \ldots, n\}$.

Solution. What makes the problem hard is that the function under consideration is very complicated; thus we try to simplify it using

inequalities. Inspired by the structure of the sum, we want to relate
$$\frac{1}{x_{i_1} + x_{i_2} + \cdots + x_{i_k}} \text{ to } \frac{1}{x_{i_1}} + \frac{1}{x_{i_2}} + \cdots + \frac{1}{x_{i_k}},$$
in order to take advantage of condition $\sum_{i=1}^{n} \frac{1}{x_i} = A$. It happens to fit the form of AM-HM inequality: $\frac{1}{a_1} + \frac{1}{a_2} + \cdots + \frac{1}{a_k} \geqslant \frac{k^2}{a_1 + a_2 + \cdots + a_k}$. Therefore we have

$$\frac{1}{a_1 + a_2 + \cdots + a_k} \leqslant \frac{\frac{1}{a_1} + \frac{1}{a_2} + \cdots + \frac{1}{a_k}}{k^2}.$$

Thus

$$\sum \frac{1}{x_{i_1} + x_{i_2} + \cdots + x_{i_k}} \leqslant \sum \frac{\frac{1}{x_{i_1}} + \frac{1}{x_{i_2}} + \cdots + \frac{1}{x_{i_k}}}{k^2}$$
$$= \frac{1}{k^2} \sum \left(\frac{1}{x_{i_1}} + \frac{1}{x_{i_2}} + \cdots + \frac{1}{x_{i_k}} \right).$$

In the sum on the right-hand side, we consider the number of appearances of each $\frac{1}{x_i}$. Apparently this is equal to the number of k-element subsets of $\{1, 2, \ldots, n\}$ containing i, which is C_{n-1}^{k-1}. Therefore

$$\frac{1}{k^2} \sum \left(\frac{1}{x_{i_1}} + \frac{1}{x_{i_2}} + \cdots + \frac{1}{x_{i_k}} \right) = \frac{1}{k^2} C_{n-1}^{k-1} \sum_{i=1}^{n} \frac{1}{x_i} = \frac{A}{k^2} C_{n-1}^{k-1}.$$

Equality holds at $x_1 = x_2 = \cdots = x_n = \frac{n}{A}$.

Therefore the maximum of $\sum \dfrac{1}{x_{i_1} + x_{i_2} + \cdots + x_{i_k}}$ is $\dfrac{A}{k^2} C_{n-1}^{k-1}$.

Example 3. Let P be a point in a regular tetrahedron T with volume 1 (including the boundary). Draw four planes passing through P, such that they are parallel to the four faces of T respectively, and divide T into 14 regions. Let $f(P)$ be the sum of the volumes of the regions that are neither tetrahedra or parallelepipeds. Find the range of $f(P)$. (IMO 31 shortlist)

Solution. Let d_1, d_2, d_3, d_4 be the distances from P to the four faces of the regular tetrahedron respectively.

Let $x_i = \dfrac{d_i}{h}$ where h is the height of the regular tetrahedron; then $\sum_{i=1}^{4} x_i = 1$.

In the 14 regions, there are four tetrahedra with volumes x_i^3, respectively. In addition, there are four parallelepipeds with volumes $6 \prod\limits_{\substack{j \ne i \\ 1 \le j \le 4}} x_j$, respectively. For example, there is a parallelepiped with three edges started from A; by symmetry there are four of them. Then

$$f(P) = 1 - \sum_{i=1}^{4} x_i^3 - 6 \sum_{1 \le i < j < k \le 4} x_i x_j x_k.$$

Obviously, $f(P) \ge 0$.

Next, we may assume that $x_1 + x_2 \le \dfrac{1}{2}$. Let $x_1 + x_2 = t \le \dfrac{1}{2}$, $x_1 x_2 = u \ge 0$, $x_3 x_4 = v \ge 0$. Since $\sum_{i=1}^{4} x_i = 1$, we have

$$\sum_{i=1}^{4} x_i^3 = (t^3 - 3tu) + (1 - t)^3 - 3(1 - t)v,$$

$$\sum_{1 \le i < j < k \le 4} x_i x_j x_k = (1 - t)u + tv.$$

Therefore

$$\begin{aligned} 1 - f(P) &= 1 - 3t + 3t^2 + 3(2 - 3t)u + 3(3t - 1)v \\ &\ge 1 - 3t + 3t^2 + 3(3t - 1)v. \end{aligned}$$

(1) If $\dfrac{1}{3} < t \le \dfrac{1}{2}$, then $3t - 1 \ge 0$, $1 - f(P) \ge 1 - 3t + 3t^2 \ge \dfrac{1}{4}$, and equality holds when $t = \dfrac{1}{2}$ and $u = v = 0$, i.e., P is the midpoint of some edge.

(2) If $0 < t \le \dfrac{1}{3}$, then $3t - 1 \le 0$, and $v = x_3 x_4 \le \dfrac{(x_3 + x_4)^2}{4} = \dfrac{(1 - t)^2}{4}$, therefore

$$1 - f(P) \geqslant 1 - 3t + 3t^2 + 3(3t - 1) \frac{(1-t)^2}{4}$$

$$= \frac{3(3t^2 + 1 - 3t)t}{4} + \frac{1}{4} \geqslant \frac{1}{4}.$$

Thus, in either case, we have $0 \leqslant f(P) \leqslant \frac{3}{4}$.

When P is a vertex of the tetrahedron, $f(P) = 0$; when P is the midpoint of any edge of the tetrahedron, $f(P) = \frac{3}{4}$.

In summary, the range of $f(P)$ is $0 \leqslant f(P) \leqslant \frac{3}{4}$.

Example 4. Assume that $a_1, a_2, \ldots, a_6; b_1, b_2, \ldots, b_6$ and $c_1, c_2, \ldots, c_6$ are three permutations of $1, 2, \ldots, 6$. Find the minimum value of $\sum_{i=1}^{6} a_i b_i c_i$. (A question in Chinese national team selection exam in 2005)

Solution. Let $S = \sum_{i=1}^{6} a_i b_i c_i$. By AM-GM inequality, we have

$$S \geqslant 6 \sqrt[6]{\prod_{i=1}^{6} a_i b_i c_i} = 6 \sqrt[6]{(6!)^3} = 6 \sqrt{6!} = 72\sqrt{5} > 160.$$

Next we prove that $S > 161$.

Because the geometric mean of the six numbers $a_1 b_1 c_1, a_2 b_2 c_2, \ldots, a_6 b_6 c_6$ is $12\sqrt{5}$, and $26 < 12\sqrt{5} < 27$, we know that one of $a_1 b_1 c_1, a_2 b_2 c_2, \ldots, a_6 b_6 c_6$ must be at least 27, and one of them must be at most 26. Since 26 is not the product of three numbers (repeatable) in $1, 2, 3, 4, 5, 6$, we know that one of them is no more than 25.

We may assume that $a_1 b_1 c_1 \geqslant 27$, $a_2 b_2 c_2 \leqslant 25$, then

$$S = (\sqrt{a_1 b_1 c_1} - \sqrt{a_2 b_2 c_2})^2 + 2\sqrt{a_1 b_1 c_1 a_2 b_2 c_2} + (a_3 b_3 c_3 + a_4 b_4 c_4)$$
$$+ (a_5 b_5 c_5 + a_6 b_6 c_6)$$
$$\geqslant (\sqrt{27} - \sqrt{25})^2 + 2\sqrt{a_1 b_1 c_1 a_2 b_2 c_2} + 2\sqrt{a_3 b_3 c_3 a_4 b_4 c_4}$$
$$+ 2\sqrt{a_5 b_5 c_5 a_6 b_6 c_6}$$

$$\geq (3\sqrt{3} - 5)^2 + 2 \cdot 3 \sqrt[6]{\prod_{i=1}^{6} a_i b_i c_i}$$

$$= (3\sqrt{3} - 5)^2 + 72\sqrt{5} > 161;$$

consequently, $S \geq 162$.

When $a_1, a_2, \ldots, a_6$; $b_1, b_2, \ldots, b_6$ and $c_1, c_2, \ldots, c_6$ are 1, 2, 3, 4, 5, 6; 5, 4, 3, 6, 1, 2 and 5, 4, 3, 1, 6, 2, respectively, we have

$$S = 1 \times 5 \times 5 + 2 \times 4 \times 4 + 3 \times 3 \times 3 + 4 \times 6 \times 1 + 5 \times 1 \times 6 + 6 \times 2 \times 2$$
$$= 162.$$

Therefore, the minimum value of S is 162.

Example 5. Given an integer $n \geq 3$ and a set of real numbers a_1, $a_2, \ldots, a_n$, satisfying $\min_{1 \leq i < j \leq n} |a_i - a_j| = 1$. Find the minimum value of $\sum_{k=1}^{n} |a_k|^3$. (CMO 2009)

Solution. We may assume that $a_1 < a_2 < \cdots < a_n$. For $1 \leq k \leq n$, we have

$$|a_k| + |a_{n-k+1}| \geq |a_{n-k+1} - a_k| \geq |n + 1 - 2k|.$$

Therefore,

$$\sum_{k=1}^{n} |a_k|^3 = \frac{1}{2} \sum_{k=1}^{n} \left(|a_k|^3 + |a_{n+1-k}|^3 \right)$$

$$= \frac{1}{2} \sum_{k=1}^{n} (|a_k| + |a_{n+1-k}|) \left(\frac{3}{4} (|a_k| - |a_{n+1-k}|)^2 \right.$$

$$\left. + \frac{1}{4} (|a_k| + |a_{n+1-k}|)^2 \right)$$

$$\geq \frac{1}{8} \sum_{k=1}^{n} \left(|a_k| + |a_{n+1-k}| \right)^3$$

$$\geq \frac{1}{8} \sum_{k=1}^{n} |n + 1 - 2k|^3.$$

When n is odd,

$$\sum_{k=1}^{n} \mid n+1-2k \mid^3 = 2 \cdot 2^3 \cdot \sum_{i=1}^{\frac{n-1}{2}} i^3 = \frac{1}{4}(n^2-1)^2.$$

When n is even,

$$\sum_{k=1}^{n} \mid n+1-2k \mid^3 = 2\sum_{i=1}^{n/2}(2i-1)^3$$

$$= 2\Big(\sum_{j=1}^{n}j^3 - \sum_{i=1}^{n/2}(2i)^3\Big)$$

$$= \frac{1}{4}n^2(n^2-2).$$

Therefore, when n is odd,

$$\sum_{k=1}^{n} \mid a_k \mid^3 \geqslant \frac{1}{32}(n^2-1)^2;$$

When n is even,

$$\sum_{k=1}^{n} \mid a_k \mid^3 \geqslant \frac{1}{32}n^2(n^2-2).$$

Equality holds when $a_i = i - \dfrac{n+1}{2}$, $i = 1, 2, \ldots, n$.

Therefore, the minimum value of $\sum_{k=1}^{n} \mid a_k \mid^3$ is $\dfrac{1}{32}(n^2-1)^2$

(n is odd), or $\dfrac{1}{32}n^2(n^2-2)$ (n is even).

Example 6. For a positive integer M, if there are integers a, b, c, d such that $M \leqslant a < b \leqslant c < d \leqslant M+49$, and $ad = bc$, we say that M is a good number; otherwise M is called a bad number. Find the biggest good number and the smallest bad number. (China TST 2006) **Solution.** The biggest good number is 576, the smallest bad number is 443.

First we prove that M is good if and only if there exist positive integers u, v, such that $uv \geqslant M$, $(u+1)(v+1) \leqslant M+49$ (*)

To prove this, first assume that u, v exist. We may assume $u \leqslant v$, thus $M \leqslant uv < u(v+1) \leqslant v(u+1) < (u+1)(v+1) \leqslant M+49$, so we can set $a = uv$, $b = u(v+1)$, $c = v(u+1)$, $d = (u+1)(v+1)$ and the requirements are satisfied (obviously $ad = bc$).

On the other hand, assume a, b, c, d exist. Since $ad = bc$ we have $\frac{a}{b} = \frac{c}{d}$; let the reduced form of this fraction be $\frac{u}{r}$; we then have $a = uv$, $b = rv$, $c = us$, $d = rs$. From $a < b$ we have $u < r$, $r \geqslant u + 1$; from $a < c$ we have $v < s$, $s \geqslant v+1$.

Therefore $uv = a \geqslant M$, $(u+1)(v+1) \leqslant rs = d \leqslant M+49$.

Now we can find the biggest good number. If M is good, then from $(*)$ and by Cauchy-Schwartz inequality, we have

$$\sqrt{M+49} \geqslant \sqrt{(u+1)(v+1)} \geqslant \sqrt{uv} + 1 \geqslant \sqrt{M} + 1.$$

So $M \leqslant 576$. When $M = 576$, we can choose $u = v = 24$ and the requirements are satisfied; so the biggest good number is 576.

Next we find the smallest bad number. First we prove that 443 is bad.

In fact, assuming that there exist u, v satisfying $(*)$ with $M = 443$, then

$$\sqrt{492} \geqslant \sqrt{(u+1)(v+1)} \geqslant \sqrt{uv} + 1,$$

then $uv \leqslant (\sqrt{492} - 1)^2 = 493 - 2\sqrt{492} < 493 - 2\sqrt{484} = 449$, thus $443 \leqslant uv \leqslant 448$.

Note that when the product uv is fixed, it is well-known that when u and v get closer, $u+v$ becomes smaller, and $(u+1)(v+1) = uv + u + v + 1$ also becomes smaller.

Now,

if $uv = 443 = 1 \cdot 443$, then $u+v \geqslant 444$, $(u+1)(v+1) \geqslant 888$;

If $uv = 444 = 12 \cdot 37$, then $u+v \geqslant 49$, $(u+1)(v+1) \geqslant 494$;

If $uv = 445 = 5 \cdot 89$, then $u+v \geqslant 94$, $(u+1)(v+1) \geqslant 540$;

If $uv = 446 = 2 \cdot 223$, then $u+v \geqslant 225$, $(u+1)(v+1) \geqslant 672$;

If $uv = 447 = 3 \cdot 149$, then $u+v \geqslant 152$, $(u+1)(v+1) \geqslant 600$;

If $uv = 448 = 16 \cdot 28$, then $u+v \geqslant 44$, $(u+1)(v+1) \geqslant 493$.

Each of these contradicts the condition $(u+1)(v+1) \leqslant 492$, so 443 is indeed a bad number.

Next we prove that any positive integer less than 443 is good.

When M belongs to one of the following intervals: $[245, 258]$; $[259, 265]$; $[266, 274]$; $[275, 280]$; $[281, 292]$; $[293, 300]$; $[301, 311]$; $[312, 322]$; $[323, 328]$; $[329, 334]$; $[335, 341]$; $[342, 350]$; $[351, 358]$; $[359, 366]$; $[367, 375]$; $[376, 382]$; $[383, 385]$; $[386, 391]$; $[392, 400]$; $[401, 406]$; $[407, 412]$; $[413, 418]$; $[419, 425]$; $[426, 430]$; $[413, 433]$; $[434, 436]$; $[437, 442]$, we can choose the corresponding (u, v) to be $(13, 20)$; $(13, 21)$; $(14, 20)$; $(17, 17)$; $(14, 21)$; $(17, 18)$; $(13, 24)$; $(18, 18)$; $(11, 30)$; $(17, 20)$; $(11, 31)$; $(16, 22)$; $(19, 19)$; $(16, 23)$; $(15, 25)$; $(16, 24)$; $(15, 26)$; $(14, 28)$; $(20, 20)$; $(17, 24)$; $(18, 23)$; $(20, 21)$; $(17, 25)$; $(18, 24)$; $(14, 31)$; $(20, 22)$; $(17, 26)$.

Finally, for $1 \leqslant M \leqslant 245$, assume $t^2 \leqslant M < (t+1)^2$, then $1 \leqslant t \leqslant 15$. If $t^2 \leqslant M < t(t+1)$, then we set $u = t$, $v = t+1$, so that $uv \geqslant M$, $(u+1)(v+1) - M \leqslant (t+1)(t+2) - t^2 = 3t+2 \leqslant 47$. If $t(t+1) \leqslant M < (t+1)^2$, then we set $u = v = t+1$, so that $uv \geqslant M$, $(u+1)(v+1) - M \leqslant (t+2)^2 - t(t+1) = 3t+4 \leqslant 49$.

Therefore, the biggest good number is 576, and the smallest bad number is 443.

Exercise 1

(1) Assume a, b, c, $a+b-c$, $b+c-a$, $c+a-b$, $a+b+c$ are seven different prime numbers, such that sum two of a, b and c add up to 800. Assume d is the difference between the maximum and minimum of these seven primes. Find the maximum value of d. (CMO 2001)

(2) Assume there are $2n$ real numbers a_1, a_2, $\ldots$, a_{2n}, satisfying $\sum_{i=1}^{2n-1} (a_{i+1} - a_i)^2 = 1$. Find the maximum value of $(a_{n+1} + a_{n+2} + \cdots +$

$a_{2n}) - (a_1 + a_2 + \cdots + a_n)$. (CWMO 2003)

(3) Assume that a_1, a_2, $\ldots$, a_n is a permutation of 1, 2, $\ldots$, n. Find the maximum value of $S_n = |a_1 - 1| + |a_2 - 2| + \cdots + |a_n - n|$.

(4) Assume that x_k ($k = 1, 2, \ldots, 1991$) satisfy $|x_1 - x_2| + |x_2 - x_3| + \cdots + |x_{1990} - x_{1991}| = 1991$. Let $y_k = \dfrac{x_1 + x_2 + \cdots + x_k}{k}$ ($k = 1$, 2, $\ldots$, 1991). Find the maximum value of $F = |y_1 - y_2| + |y_2 - y_3| + \cdots + |y_{1990} - y_{1991}|$. (The 25th Russian Mathematical Olympiad)

(5) Assume that x_1, x_2, $\ldots$, x_{1990} is a permutation of 1, 2, $\ldots$, 1990. Find the maximum value of

$$F = |\cdots||x_1 - x_2| - x_3| - \cdots - x_{1990}|.$$

(The 24th Russian Mathematical Olympiad)

(6) Assume $0 < p \leqslant a_i \leqslant q$, b_i is a permutation of a_i ($1 \leqslant i \leqslant n$). Find the maximum value of $F = \sum_{i=1}^{n} \dfrac{a_i}{b_i}$. (Hungary Mathematical Olympiad)

Chapter 2   Repeated Extremum

One of the characteristics of Combinatorial Extremization is the relatively large number of variables in the target function, which makes it difficult to find the trend of the function. If we freeze some of the variables, i.e. set these variables constant, then the relation between the target function and the remaining variables will be more evident, thus the first extremum can be relatively easily obtained. Afterwards we "*unfreeze*" the frozen variables and then find the extremum of the function.

Typically there are two approaches to freezing variables. The first one is to freeze only one variable, which is normally used for finding extremum of a ternary function: For the ternary function $f(x, y, z)$, if z is fixed, then the function can be seen as a binary function of x, y. Based on this we determine the extremum $G(z)$ of the binary function, and then treat z as a variable to find the extremum of $G(z)$. Its basic idea is as follows:

$$u = f(x, y, z) = g(x, y) \leqslant G(z) \leqslant C.$$

However, in some cases, the expression of $G(z)$ is a piecewise function, thus the above idea can be expressed as

$$u = f(x, y, z) = g(x, y) \leqslant G(z) = \begin{cases} G_1(z) & (z \in A) \\ G_2(z) & (z \in B) \end{cases}$$

$$\leqslant \begin{cases} A_1 & (z \in A) \\ A_2 & (z \in B) \end{cases} \Rightarrow u \leqslant \max\{A_1, A_2\}.$$

In particular, if the equalities in $g(X, Y) \leqslant G(z) \leqslant C$ do not hold simultaneously, we must consider different scenarios with respect to the fixed value of z and deal with them respectively (some specific values of z will be discussed separately). The basic idea is shown below:

$$u = f(x, y, z) = \begin{cases} g_1(x, y) & (z = z_1) \\ \vdots & \vdots \\ g_k(x, y) & (z = z_k) \\ g(x, y) & (z \in A) \end{cases} \leqslant \begin{cases} G_1 & (z = z_1) \\ \vdots & \vdots \\ G_k & (z = z_k) \\ G(z) \leqslant G & (z \in A) \end{cases}$$

$\Rightarrow u \leqslant G$, where $G = \max\{G_1, G_2, \ldots, G_k, G_A\}$.

The second approach is to freeze multiple variables, which is usually used to find the extremum of multivariable functions (those involving > 3 variables): in the analytic expression of the multivariable function, select one of the variables as the main argument, freeze all other variables, so that the function becomes a single variable function $f(t)$. We first obtain the extremum $f(t_0)$ of $f(t)$, and then freeze variables again for $f(t_0)$ (since $f(t_0)$ is a function of the other variables), turning it into a single variable function to find the extremum. This process goes on until the extremum of the function is obtained.

In essence, repeated extremum is precisely enlarging and shrinking, which eliminate the variables step by step by fixing the variables.

Let us see an example of finding the extremum of a general function.

Example 1. Let x, y, z be nonnegative real numbers $x + y + z = 1$. Find the extremum of $F = 2x^2 + y + 3z^2$.

Solution. First, by substitution, we have $y = 1 - x - z$, thus

$$F = 2x^2 + 1 - x - z + 3z^2$$
$$= 2\left(x - \frac{1}{4}\right)^2 + 3\left(z - \frac{1}{6}\right)^2 + \frac{19}{24}$$
$$\geqslant \frac{19}{24}.$$

and $F\left(\dfrac{1}{4}, \dfrac{7}{12}, \dfrac{1}{6}\right) = \dfrac{19}{24}$, thus the minimum value of F is $\dfrac{19}{24}$.

Now we find the maximum value of F by repeated extremization. Fix the variable z, and then $x + y = 1 - z$ (constant).

For $F = 2x^2 + y + 3z^2$, since z is a constant, we only have to find the maximum value of $2x^2 + y = A$, where $x + y = 1 - z$. For convenience, let $1 - z = t$, then $x + y = t$, $0 \leqslant x$, $y \leqslant t \leqslant 1$, t is a constant.

Since $A = 2x^2 + y = 2x^2 + t - x$ (substitution), notice that $0 \leqslant x \leqslant t \leqslant 1$, and that the graph of this quadratic function opens upwards, the maximum value of the function is not attained at the vertex. Thus the maximum value of A can only be reached at $x = 0$ or $x = t$. Therefore,

$$g(z) = A_{\max} = \max\{t, 2t^2\} = \begin{cases} t & \left(0 \leqslant t \leqslant \dfrac{1}{2}\right), \\ 2t^2 & \left(\dfrac{1}{2} \leqslant t \leqslant 1\right). \end{cases}$$

Equivalently

$$g(z) = A_{\max} = \max\{1 - z, 2(1 - z)^2\} = \begin{cases} 1 - z & \left(\dfrac{1}{2} \leqslant z \leqslant 1\right), \\ 2(1 - z)^2 & \left(0 \leqslant z \leqslant \dfrac{1}{2}\right). \end{cases}$$

Now we find the maximum value of $g(z)$.

$F \leqslant g(z) + 3z^2 = 2(1 - z)^2 + 3z^2 = 5z^2 - 4z + 2 \leqslant 2$, when $0 \leqslant z \leqslant \dfrac{1}{2}$.

$F \leqslant g(z) + 3z^2 = (1 - z) + 3z^2 \leqslant 3$, when $\dfrac{1}{2} \leqslant z \leqslant 1$.

This means that, for all x, y and z, we have $F \leqslant 3$, while equality holds at $x = y = 1$, and $z = 0$. Therefore, the maximum value of F is 3.

In summary, the minimum value of F is $\dfrac{19}{24}$, and the maximum value of F is 3.

Note that it is rather simple to find the maximum value of F by enlarging and shrinking the power and coefficients of F. Actually, $2x^2 + y + 3z^2 \leqslant 2x + y + 3z \leqslant 3x + 3y + 3z \leqslant 3$.

Example 2. Assume there are 1988 unit cubes. Put them together to construct three regular square prisms A, B, C with height 1 and bottom side lengths a, b, c respectively. Put A, B and C in the first quadrant, so that the bottom sides are parallel to the axes. Assume that one vertex of C is the origin. B is placed on top of C, and each unit cube of B lies on top of exactly one unit cube of C, but the boundary of B does not touch the boundary of C. Similarly, A is placed on top of B, each unit cube of A lies on top of exactly one unit cube of B, but the boundary of A does not touch the boundary of B. A three-storey building is built like this. What is the value of a, b and c, such that the number of different buildings is maximized? (11th math competition of Australia-Poland)

Analysis and Solution. From the assumption that the boundaries do not match, we have $a \leqslant b - 2 \leqslant c - 4$. Then, there are $(b - a - 1)^2$ ways to put A on top of B, and $(c - b - 1)^2$ ways to put B on C. Then we have $P = (b - a - 1)^2 (c - b - 1)^2$ different three-storey buildings. The question is then reduced to finding the maximum of $P = (b - a - 1)^2 (c - b - 1)^2$ where a, b, c are three positive integers satisfying $1 \leqslant a \leqslant b - 2 \leqslant c - 4$, $a^2 + b^2 + c^2 \leqslant 1988$.

Obviously, $P \leqslant (b - 2)^2 (c - b - 1)^2$ and equality holds when $a = 1$. So we have to find the maximum value of $Q = (b - 2)(c - b - 1)$ where the positive integers b, c satisfy $3 \leqslant b \leqslant c - 2$, $b^2 + c^2 \leqslant 1987$.

One naive way is to estimate

$$Q \leqslant \frac{[(b-2)+(c-b-1)]^2}{4} = \frac{(c-3)^2}{4}. \qquad (*)$$

Now we only have to find the value of c which maximizes $\frac{(c-3)^2}{4}$. From the conditions we have $c^2 \leqslant 1987 - b^2 \leqslant 1987$, $c \leqslant 44$, thus $Q \leqslant \frac{(44-3)^2}{4} < 421$, $Q \leqslant 420$.

Unfortunately, equality does not hold here. So we have to use the repeated extremization method.

Fix c, we have

$$Q = -b^2 + (1+c)b + 2 - 2c$$
$$= -\left(b - \frac{1+c}{2}\right)^2 + \frac{(1+c)^2}{4} + 2 - 2c. \qquad (**)$$

From the graph of the quadratic function, we have that $Q \leqslant \frac{(1+c)^2}{4} + 2 - 2c = A(c)$. (When c is odd, $b = \frac{1+c}{2}$.) Moreover $Q \leqslant \frac{(1+c)^2}{4} - \frac{1}{4} + 2 - 2c = B(c)$. (When c is even, $b = \frac{c}{2}$.)

Now we find the maximum value of $A(c)$ and $B(c)$.

Since $b^2 + c^2 \leqslant 1987$, we have $c \leqslant 44$. Thus when c is odd,

$$Q \leqslant A(c) \leqslant A(43) = \frac{(1+43)^2}{4} + 2 - 86 = 400. \qquad ①$$

When c is even,

$$Q \leqslant B(c) \leqslant B(44) = \frac{(1+44)^2}{4} - \frac{1}{4} + 2 - 88 = 420. \qquad ②$$

However, equality still cannot hold. Actually, equality holds in ② when $c = 44$, and $b = \frac{c}{2} = 22$. However $(44, 22)$ does not satisfy $b^2 + c^2 \leqslant 1987$. Therefore the maximum value of $Q = (b-2)(c-b-1)$ cannot be found by fixing c. We should discuss the problem according to the value of c, and use different forms of target functions.

When $c = 44$, $Q = -b^2 + 45b - 86$. Since $b^2 \leqslant 1987 - c^2 = 51$, we know $3 \leqslant b \leqslant 7$. So when $b = 7$, Q has the maximum value 180.

When $c = 43$, $3 \leqslant b \leqslant 11$. Now $Q \leqslant 324$ and equality holds when $b = 11$.

When $c = 42$, $3 \leqslant b \leqslant 14$. Now $Q \leqslant 324$ and equality holds when $b = 14$.

When $c = 41$, $3 \leqslant b \leqslant 17$. Now $Q \leqslant 345$ and equality holds when $b = 17$.

When $c = 40$, $3 \leqslant b \leqslant 19$. Now $Q \leqslant 320$ and equality holds when $b = 19$.

When $c \leqslant 39$,

$$Q \leqslant \frac{[(b-2) + (c-b-1)]^2}{4} = \frac{(c-3)^2}{4} \leqslant 18^2 = 324.$$

As a result, $Q \leqslant 345$ always holds, and when $(a, b, c) = (1, 17, 41)$ equality holds. So when $(a, b, c) = (1, 17, 41)$ the maximum value of P is 345^2.

Example 3. Given a positive integer k and a positive real number a. For any partition $k_1 + k_2 + \cdots + k_r = k$ (k_i is positive integer, $1 \leqslant r \leqslant k$), find the maximum of $F = a^{k_1} + a^{k_2} + \cdots + a^{k_r}$ (8th China Mathematical Olympiad)

Solution. The essence of the problem is to divide k into some positive integers k_i, such that the value of $a^{k_1} + a^{k_2} + \cdots + a^{k_r}$ is maximized. Since the number of terms is not fixed, we should proceed in two steps (repeated extremization). First we fix r, assume that k is divided into r positive integers $k_i (i = 1, 2, \ldots, r)$, and find the maximum value of $a^{k_1} + a^{k_2} + \cdots + a^{k_r}$. Then let r vary, and find the maximum value of $f(r)$.

In the first step, consider for example the case $k = 6$, $r = 3$, then $k_1 + k_2 + k_3 = 6$, $F = a^{k_1} + a^{k_2} + a^{k_3}$.

(1) If $(k_1, k_2, k_3) = (2, 2, 2)$, then $F_1 = a^2 + a^2 + a^2 = 3a^2$.

(2) If $(k_1, k_2, k_3) = (1, 2, 3)$, then $F_2 = a + a^2 + a^3$.

(3) If $(k_1, k_2, k_3) = (1, 1, 4)$, then $F_3 = a + a + a^4 = 2a + a^4$.

Now $F_2 - F_1 = a - 2a^2 + a^3 = a(1 - 2a + a^2) = a(1-a)^2 \geqslant 0$,

$$F_3 - F_2 = a + a^4 - a^2 - a^3 = a(1 + a^3 - a^2 - a)$$
$$= a(1 - a^2)(1 - a) \geqslant 0.$$

So F_3 is the largest.

In general it is not difficult to guess that, when $(k_1, k_2, \ldots, k_r)$ is in some sense *"concentrated"*, F will achieve its maximum. That is, the point of maximum of F is $(1, 1, \ldots, k-r+1)$. We first prove the following lemma.

Lemma. Set $a > 0$, x, $y \in \mathbf{N}^*$. Then $a^x + a^y \leqslant a^{x+y-1} + a$.

Proof. $a^{x+y-1} + a - a^x - a^y = a[a^{x-1} - 1][a^{y-1} - 1] \geqslant 0$. □

Use the lemma repeatedly, we have

$$
\begin{aligned}
F &= a^{k_1} + a^{k_2} + \cdots + a^{k_r} \\
&\leqslant a + a^{k_1+k_2-1} + a^{k_3} + \cdots + a^{k_r} \\
&\leqslant a + a + a^{k_1+k_2+k_3-2} + a^{k_4} + \cdots + a^{k_r} \\
&\leqslant \cdots \leqslant a + a + \cdots + a + a^{k_1+k_2+\cdots+k_r-(r-1)} \\
&= (r-1)a + a^{k-r+1}.
\end{aligned}
$$

Next we find the maximum value of $f(r) = (r-1)a + a^{k-r+1}$ for $1 \leqslant r \leqslant k$.

Let $f(x) = a(x-1) + a^{k-x+1}$, since $a(x-1)$ and a^{k-x+1} are both convex functions, $f(x)$ is also a convex function. Thus

$$f(r) \leqslant \max\{f(1), f(k)\} = \max\{a^k, ka\}.$$

In summary, the maximum value of F is $\max\{a^k, ka\}$.

Example 4. Consider a quadrilateral $ABCD$ inscribed in a circle, whose four sides have lengths being positive integers. We know $DA = 2005$, $\angle ABC = \angle ADC = 90°$ and $\max\{AB, BC, CD\} < 2005$. Find the maximum and minimum value of the perimeter of quadrilateral $ABCD$. (National Training Team of China in 2005)

Solution. Set $AB = a$, $BC = b$, $CD = c$, then $a^2 + b^2 = AC^2 = c^2 + 2005^2$. Thus $2005^2 - a^2 = b^2 - c^2 = (b+c)(b-c)$, where $a, b, c \in \{1, 2, \ldots, 2004\}$.

Let $a \geqslant b$, fix a, set $a_1 = 2005 - a$, then

$$(b+c)(b-c) = 2005^2 - a^2 = a_1(4010 - a_1). \qquad ①$$

From $a^2 + b^2 > 2005^2$, we have $a > \dfrac{2005}{\sqrt{2}} > 1411$, thus

$$1 \leqslant a_1 < 2005 - 1411 = 594.$$

From ①, we have $b + c > \sqrt{(b+c)(b-c)} = \sqrt{a_1(4010 - a_1)}$.

When $a_1 = 1$, $a = 2004$, we have $(b+c)(b-c) = 4009 = 19 \times 211$, thus $b + c \geqslant 211$, $a + b + c \geqslant 2004 + 211 = 2215 > 2155$.

When $a_1 = 2$, $a = 2003$, we have

$$(b+c)(b-c) = 2^4 \times 3 \times 167.$$

Since $b + c$ and $b - c$ have the same parity, we have $b + c \geqslant 2 \times 167$, $a + b + c \geqslant 2003 + 2 \times 167 > 2155$.

When $a_1 = 3$, $a = 2002$, we have $(b+c)(b-c) = 3 \times 4007$. Thus $b + c \geqslant 4007$, $a + b + c \geqslant 2002 + 4007 > 2155$.

When $a_1 = 4$, $a = 2003$, we have $(b+c)(b-c) = 2^3 \times 2003$, thus $b + c \geqslant 2 \times 2003$, $a + b + c \geqslant 2001 + 2 \times 2003 > 2155$.

When $a_1 = 5$, $a = 2000$, we have $(b+c)(b-c) = 3^2 \times 5^2 \times 89$, thus $b + c \geqslant 225$, $a + b + c \geqslant 2000 + 225 > 2155$.

When $a_1 = 6$, $a = 1999$, we have $(b+c)(b-c) = 6 \times 4004 = 156 \times 154$, thus $b + c \geqslant 156$, $a + b + c \geqslant 1999 + 156 = 2155$.

When $a_1 \geqslant 7$, since $b + c > \sqrt{a_1(4010 - a_1)}$, we have

$$a + b + c > \sqrt{a_1(4010 - a_1)} + 2005 - a_1.$$

However, $7 \leqslant a_1 < 594$, and we have that $\sqrt{a_1(4010 - a_1)} + 2005 - a_1 > 2155$.

In fact,

$$\sqrt{a_1(4010 - a_1)} + 2005 - a_1 > 2155 \Leftrightarrow \sqrt{a_1(4010 - a_1)} > 150 + a_1$$
$$\Leftrightarrow -a_1^2 + 4010a_1 > a_1^2 + 300a_1 + 150^2 \Leftrightarrow a_1^2 - 1855a_1 + 11250 < 0.$$

From the properties of the quadratic functions, the inequality holds when $7 \leqslant a_1 < 594$.

In conclusion, we have $a + b + c \geqslant 2155$. Thus

$$AB + BC + CD + DA \geqslant 2155 + 2005 = 4160.$$

When $AB = 1999$, $BC = 155$, $CD = 1$, we can check that equality holds. So the minimum value of the perimeter of the quadrilateral is 4160.

Next we find the maximum value of the perimeter of quadrilateral $ABCD$.

Since $a \geqslant b$, $c < 2005$, $b + c < a + 2005 = 4010 - a_1$. From expression ①, we know that $a_1 < b - c < b + c < 4010 - a_1$.

Since a_1 and $b - c$ have the same parity, $b - c \geqslant a_1 + 2$. Thus

$$b + c = \frac{a_1(4010 - a_1)}{b - c} \leqslant \frac{a_1(4010 - a_1)}{a_1 + 2}.$$

When $b - c = a_1 + 2$,

$$a + b + c = 2005 - a_1 + \frac{a_1(4010 - a_1)}{a_1 + 2}$$

$$= 6021 - 2\left(a_1 + 2 + \frac{4012}{a_1 + 2}\right).$$

But $4012 = 2^2 \times 17 \times 59 = 68 \times 59$, so $a + b + c \leqslant 6021 - 2 \cdot (68 + 59) = 5767$.

When $b - c \neq a_1 + 2$, we have $b - c \geqslant a_1 + 4$. Then

$$a + b + c \leqslant 2005 - a_1 + \frac{a_1(4010 - a_1)}{a_1 + 4}.$$

But

$$2005 - a_1 + \frac{a_1(4010 - a_1)}{a_1 + 4} \leqslant 5767 \Leftrightarrow a_1^2 - 122a_1 + 7524 \geqslant 0.$$

Notice that $\Delta = 122^2 - 4 \times 7524 < 0$, so the above inequality is true.

In any case $AB + BC + CD + DA \leqslant 5767 + 2005 = 7772$. When $a = 1948$, $b = 1939$, $c = 1880$ we can check that equality holds, thus the maximum value of the perimeter of quadrilateral $ABCD$ is 7772.

Exercise 2

(1) Assume that x, y, z are nonnegative real numbers, and $x + y + z = a$ $(a \geqslant 1)$. Find the maximum value of $F = 2x^2 + y + 3z^2$.

(2) Find a 3-digit natural number, such that the ratio between itself and the sum of its digits is minimized.

(3) Assume that x_1, x_2, ..., x_n are nonnegative real numbers, let the maximum value of $H = \dfrac{x_1}{(1 + x_1 + x_2 + \cdots + x_n)^2} + \dfrac{x_2}{(1 + x_2 + x_3 + \cdots + x_n)^2} + \cdots + \dfrac{x_n}{(1 + x_n)^2}$ be a_{n+1}. Find the values of x_1, x_2, ..., x_n that maximize H. Also find the relation between a_n and a_{n-1} and $\lim\limits_{n \to \infty} a_n$.

(4) Assume that n is a given positive integer $(n > 1)$, consider positive integers a, b, c, d that satisfy $\dfrac{b}{a} + \dfrac{d}{c} < 1$, $b + d \leqslant n$. Find the maximum value of $\dfrac{b}{a} + \dfrac{d}{c}$.

In this approach, we first prove that the required extremum exists, then guess the extremum point by the heuristics suggested by the problem, and then prove that the function cannot reach its extremum at other points. Assume that the function reaches the maximum at another point $(x_1, x_2, \ldots, x_n)$, we will appropriately adjust the variables (usually we increase the small terms, and decrease the large terms), and show that the function takes a larger or smaller value at point $(x'_1, x'_2, \ldots, x'_n)$, thus showing that the extremum is not attained at this point. The basic steps are as follows:

Prove that the extremum exists — guess the extreme point — prove that any other point is not the extremum point — deduce the conclusion.

Example 1. Consider some positive integers whose sum is 1976. Find the maximum value of the product of these positive integers. (18th IMO)

Analysis. First, we look at the simple situations in which the sums are 4, 5, 6, 7 and 8 respectively. The partitions which maximize the product are as follows: $4 = 2+2$, $5 = 2+3$, $6 = 3+3$, $7 = 2+2+3$, $8 = 2+3+3$.

We can guess that, to maximize the product, the partition should consist of only numbers 2 and 3, and has at most two 2.

Solution. First, the number of partitions of 1976 is finite. Thus there must be a partition whose product is the biggest.

Suppose that $(x_1, x_2, \ldots, x_n)$ is a maximizing partition. We will prove that, we can arrange so that this partition satisfies the following properties:

(1) $x_i \leqslant 3$.

If some $x_i \geqslant 4$, split x_i to two numbers: 2 and $x_i - 2$. Then we have a new partition: $(x_1, x_2, \ldots, x_{i-1}, x_i - 2, 2, x_{i+1}, \ldots, x_n)$. Since $2(x_i - 2) = 2x_i - 4 \geqslant x_i$, after splitting the value of P will not decrease.

(2) $x_i \neq 1$.

If some $x_i = 1$, then choose any number x_j in the partition, and replace $x_i = 1$ and x_j with a new number $(1 + x_j)$, and thus obtain a new partition $(x_1, x_2, \ldots, x_{j-1}, x_{j+1}, \ldots, x_n, x_j + 1)$. Since $1 \cdot x_j < 1 + x_j$, after doing this P will increase.

(3) The number of 2's in the partition is no more than 2.

If $x_i = x_j = x_k = 2$, we replace x_i, x_j, x_k with two new numbers 3 and 3, and obtain a new partition. Since $2 \times 2 \times 2 < 3 \times 3$, P will increase after the operation.

Thus, x_i equals to 2 or 3, and the number of 2 is no more than 2.

Notice that $1976 = 658 \times 3 + 2$, thus the maximum value of P is $3^{658} \times 2$.

Example 2. There are 1989 points in the space, no three being collinear. Divide them into 30 groups, the number of points in each group being different. Select three points from three different groups respectively, and form a triangle using these three points as vertices. What are the numbers of elements in these groups when the number of possible triangles is maximized? (4th China Mathematics Olympiad)

Analysis. Intuitively, the number of triangles is maximized when the groups have the same number of points. However, the assumption requires that the numbers of points in the groups are different from each other. Hence we feel that the numbers of points in the groups should be sufficiently close. To justify this feeling, we now consider a special case.

First, look at the situation when 10 points are divided into three groups. When the numbers of points are 1, 2, 7 respectively, the number of triangle is $S = 14$, denoted by $S(1, 2, 7) = 14$. Similarly, $S(1, 3, 6) = 18$, $S(1, 4, 5) = 20$, $S(2, 3, 5) = 30$. Note $S(2, 3, 5) = 30$ is largest. In general, we can conjecture that S is maximized when the numbers of points n_i in the groups are close to each other. That is, the difference between any two adjacent numbers n_t and n_{t+1} should be as small as possible. Apparently, n_t and n_{t+1} should differ by at least 1. Can all the pairs n_t and n_{t+1} satisfy $n_{t+1} - n_t = 1$? This will probably lead to a way to solve the problem.

Solution. Let the numbers of points in the groups be $n_1 < n_2 < \cdots < n_{30}$. Then the number of triangles is

$$S = \sum_{1 \leqslant i < j < k \leqslant 30} n_i n_j n_k, \text{ with } n_1 + n_2 + \cdots + n_{30} = 1989.$$

Since the number of partitions is finite, S has a maximum value. Assume that $(n_1, n_2, \ldots, n_{30})$ maximizes S and $n_1 < n_2 < \cdots < n_{30}$, then we have the following:

(1) For any $t = 1, 2, \ldots, 29$, we have $n_{t+1} - n_t \leqslant 2$.

In fact, assume there exists $1 \leqslant t \leqslant 29$ in which $n_{t+1} - n_t \geqslant 3$ (or just assume $n_2 - n_1 \geqslant 3$), let $n'_t = n_t + 1$, $n'_{t+1} = n_{t+1} - 1$, then the numbers of points in each group are still different. Now

$$S = \sum_{1 \leqslant i < j < k \leqslant 30} n_i n_j n_k$$

$$= n_t n_{t+1} \cdot \sum_{\substack{k \neq t, \, t+1 \\ 1 \leqslant k \leqslant 30}} n_k + (n_t + n_{t+1}) \cdot \sum_{\substack{j, \, k \neq t, \, t+1 \\ 1 \leqslant j < k \leqslant 30}} n_j n_k + \sum_{\substack{i, \, j, \, k \neq t, \, t+1 \\ 1 \leqslant i < j < k \leqslant 30}} n_i n_j n_k.$$

$$S' = n'_t n'_{t+1} \cdot \sum_{\substack{k \neq t, \, t+1 \\ 1 \leqslant k \leqslant 30}} n_k + (n'_t + n'_{t+1}) \cdot \sum_{\substack{j, \, k \neq t, \, t+1 \\ 1 \leqslant j < k \leqslant 30}} n_j n_k + \sum_{\substack{i, \, j, \, k \neq t, \, t+1 \\ 1 \leqslant i < j < k \leqslant 30}} n_i n_j n_k.$$

Since $n'_t + n'_{t+1} = n_t + n_{t+1}$, and $n'_t n'_{t+1} = n_t n_{t+1} - n_t + n_{t+1} - 1 > n_t n_{t+1}$, we have $S' > S$, which is a contradiction. So $n_{t+1} - n_t = 1$ or 2.

(2) There is at least one $t (1 \leqslant t \leqslant 29)$, such that $n_{t+1} - n_t = 2$. In fact, if $n_{t+1} - n_t = 1$, then $n_1, n_2, \ldots, n_{30}$ are 30 consecutive positive integers, so their sum is multiple of 15. However, $\sum_{t=1}^{30} n_t = 1989$ is not a multiple of 15, which is a contradiction.

(3) There is at most one t $(1 \leqslant t \leqslant 29)$, such that $n_{t+1} - n_t = 2$.

In fact, if there are s, t $(1 \leqslant s < t \leqslant 29)$ such that $n_{t+1} - n_t = n_{s+1} - n_s = 2$, then let $n'_s = n_s + 1$, $n'_{t+1} = n_{t+1} - 1$ (we decrease the largest one and increase the smallest one). The numbers of points in the groups are still different, and

$$S = \sum_{1 \leqslant i < j < k \leqslant 30} n_i n_j n_k$$

$$= n_s n_{t+1} \cdot \sum_{\substack{k \neq s, t+1 \\ 1 \leqslant k \leqslant 30}} n_k + (n_s + n_{t+1}) \cdot \sum_{\substack{j, k \neq s, t+1 \\ 1 \leqslant j < k \leqslant 30}} n_j n_k + \sum_{\substack{i, j, k \neq t, t+1 \\ 1 \leqslant i < j < k \leqslant 30}} n_i n_j n_k ,$$

$$S' = n'_s n'_{t+1} \cdot \sum_{\substack{k \neq t, t+1 \\ 1 \leqslant k \leqslant 30}} n_k + (n'_s + n'_{t+1}) \cdot \sum_{\substack{j, k \neq t, t+1 \\ 1 \leqslant j < k \leqslant 30}} n_j n_k + \sum_{\substack{i, j, k \neq t, t+1 \\ 1 \leqslant i < j < k \leqslant 30}} n_i n_j n_k .$$

Since $n'_s + n'_{t+1} = n_s + n_{t+1}$, and $n'_s n'_{t+1} = n_s n_{t+1} - n_s + n_{t+1} - 1 > n_s n_{t+1}$, we have $S' > S$, which is a contradiction.

From (2) and (3), there is only one t $(1 \leqslant t \leqslant 29)$ in which $n_{t+1} - n_t = 2$.

And finally we prove that the sequence $(n_1, n_2, \ldots, n_{30})$ satisfying (1), (2) and (3) simultaneously is unique.

Assume that the 30 numbers are $n_1, n_1 + 1, n_1 + 2, \ldots, n_1 + t - 1, n_1 + t + 1, n_1 + t + 2, \ldots, n_1 + 30$, then $n_1 + (n_1 + 1) + (n_1 + 2) + \cdots + (n_1 + t - 1) + (n_1 + t + 1) + (n_1 + t + 2) + \cdots + (n_1 + 30) = 1989$.

Thus $(n_1 + t) + (n_1 + 1) + (n_1 + 2) + \cdots + (n_1 + t - 1) + (n_1 + t + 1) + (n_1 + t + 2) + \cdots + (n_1 + 30) = 1989 + t$, that is, $1989 + t = 30n_1 + (1 + 2 + \cdots + 30) = 30n_1 + 15 \times 31$.

So $1974 + t = 30n_1 + 15 \times 30$, $30 \mid 1974 + t$, $30 \mid 24 + t$. Since $1 \leqslant t \leqslant 29$, we have $t = 6$.

In conclusion, the numbers of points of each group are 51, 52, $\ldots$, 56, 58, 59, $\ldots$, 81.

Example 3. Let a point P start from $A(1, 1)$, move along lattice paths, and arrive at $B(m, n)$ $(m, n \in \mathbf{N}^*)$. At each step P moves to an adjacent lattice point, so that either the x-coordinate or the y-coordinate increases by 1. Find the maximum value of the sum S of the products of x- and y-coordinates of all lattice points P passes through.

Analysis and solution. Let the points P go through be $P_1 = A(1, 1)$, $P_2, \ldots, P_{m+n-1} = B(m, n)$. Also let the coordinate of P_i be (x_i, y_i), then $S = \sum_{i=1}^{m+n-1} x_i y_i$. To maximize S, intuitively, x_i, y_i should be close to each other; but there is no way to make $x_i = y_i$ for each i (otherwise P will move along the diagonal). We guess that, if values $(x_1, y_1), (x_2, y_2), \ldots, (x_{m+n-1}, y_{m+n-1})$ maximize S, then for any $x_i < m$, $y_i < n$, $| x_i - y_i | \leqslant 1$.

Assume for some i that $| x_i - y_i | > 1 (x_i < m, y_i < n)$, say $x_i - y_i > 1$. The natural idea is then to increase y_i by 1 and decrease x_i by 1. That is, $P_i(x_i, y_i)$ is changed to $P_i'(x_i - 1, y_i + 1)$ while the rest points remain the same. But does the new route still satisfy the requirements? Apparently, to satisfy the requirements, P_i should satisfy that: $P_{i-1}P_i$ is horizontal and P_iP_{i+1} is vertical. P_i may not satisfy such requirements; however, we observe that must be another point $P_t(x_t, y_t)$ satisfying the conditions. That is, the route must contain three continuous points $P_{t-1}(x_{t-1}, y_{t-1})$, $P_t(x_t, y_t)$, $P_{t+1}(x_{t+1}, y_{t+1})$, such that $P_{t-1}P_t$ is horizontal and P_tP_{t+1} is vertical, and $x_t - y_t > 1$.

Actually, if $P_{i-1}P_i$ is vertical, then choose the point whose x-coordinate is x_i and y-coordinate is minimal. Let it be $P_t(x_t, y_t)$, where $x_t = x_i$, $y_t < y_i$, then $x_t - y_t = x_i - y_t > x_i - y_i > 1$. Since $x_t > 1 + y_t > 1$, before P arrives at $P_t(x_t, y_t)$ there must exist a horizontal line in the route. Since is y_t minimal, $P_{t-1}P_t$ is a horizontal line and P_tP_{t+1} is a vertical line. Similarly, if $P_{i-1}P_i$ is a horizontal line, we choose the point whose y-coordinate is y_i and x-coordinate is maximal. Let it be $P_t(x_t, y_t)$, where $x_t > x_i$, $y_t = y_i$. Now $x_t - y_t = x_t - y_i > x_i - y_i > 1$. Since $y_i < n$, before P arrives at $P_t(x_t, y_t)$ there must be a vertical line in the route. Then since x_t is maximal, $P_t P_{t+1}$ is vertical and $P_{t-1}P_t$ is horizontal.

In summary, when there exists point $P_i(x_i, y_i)$ where $x_i < m$ and $y_i < n$, satisfying $x_i - y_i > 1$, there must be three consecutive points $P_{t-1}(x_{t-1}, y_{t-1})$, $P_t(x_t, y_t)$ and $P_{t+1}(x_{t+1}, y_{t+1})$, satisfying that $P_{t-1}P_t$ is horizontal and $P_t P_{t+1}$ is vertical, and $x_t - y_t > 1$. Thus we replace $P_t(x_t, y_t)$ with $P_t'(x_t - 1, y_t + 1)$, so that the adjusted route still satisfies the condition, and $(x_t - 1)(y_t + 1) = x_t y_t + x_t - y_t - 1 > x_t y_t$. So after adjustment S will increase, which is a contradiction.

By the above discussions we know that, for any point $P_i(x_i, y_i)$ in the route, if $x_i \neq y_i$, then the route starting from $P_i(x_i, y_i)$ is unique, and the next point is obtained via increasing the smaller coordinate of $P_i(x_i, y_i)$ by 1. When $x_i = y_i$, there are two options to start from $P_i(x_i, y_i)$, the next point will be obtained via increasing x_i or y_i by 1. Thus, when $m \geqslant n$, the route is:

$A(1, 1) \rightarrow P_2(2, 1)$ or $P_2(1, 2) \rightarrow P_3(2, 2) \rightarrow P_4(2, 3)$ or $P_4(3, 2) \rightarrow P_5(3, 3) \rightarrow \cdots \rightarrow P_{2n-1}(n, n) \rightarrow P_{2n}(n + 1, n) \rightarrow P_{2n+1}(n + 2, n) \rightarrow \cdots \rightarrow P_{m+n-1}(m, n)$.

Thus,

$$S_{\max} = \sum_{i=1}^{n} i^2 + \sum_{i=1}^{n-1} i(i + 1) + n \sum_{i=1}^{m-n} (n + i)$$
$$= \frac{1}{6} n(3m^2 + n^2 + 3m - 1).$$

When $m < n$, similarly, $S_{\max} = \frac{1}{6} m(3n^2 + m^2 + 3n - 1)$.

Example 4. The MO space station consists of 99 space stations, where any two stations are connected by a tubular channel. Set 99 of the channels to be two-way channels, and the rest channels are strictly one way. For a group of four stations, if starting from any station one can reach any other station through the channels, the group of four stations is called a connected four-station group.

Find the maximum number of the connected four-station groups, and justify your answer. (CMO1999)

Analysis and Solution. It is rather difficult to find the number of connected four-station groups straightforward because the conditions are rather strong, but it is relatively easy to satisfy the condition for unconnected four-station groups. For example, one may consider an MO station with no "*circuit*". This leads to the idea of considering all one-way channels starting from the same point, where any 3-channel triple corresponds to an unconnected four-station group. Of course there may be other unconnected groups, but those unconnected groups do not necessarily exist. That is to say, there may be a plan such that the number of those unnecessarily existing groups is zero, so we do not need to consider them.

Let the 99 stations be A_1, A_2, ..., A_{99}. Let the number of one-way channels starting from A_i, the number of one-way channels ending at A_i and the number of two-way channels passing through A_i be w_i, l_i and k_i, respectively. From the conditions, we have $w_i + l_i + k_i = 98$, and $k_1 + \cdots + k_{99} = 198$, $w_1 + \cdots + w_{99} = l_1 + \cdots + l_{99}$, then

$$\sum_{i=1}^{99} w_i = \sum_{i=1}^{99} l_i = \frac{1}{2} \sum_{i=1}^{99} (w_i + l_i)$$

$$= \frac{1}{2} \sum_{i=1}^{99} (w_i + l_i + k_i) - \frac{1}{2} \sum_{i=1}^{99} k_i = 4752.$$

Note that there are w_i one-way channels going from A_i; for any three of them, we obtain a four-station group, which is easily seen to be unconnected. Since any unconnected group contains at most one station A with three outgoing channels, we know the number of unconnected groups $S \geqslant \sum_{i=1}^{99} C_{w_i}^3$.

Now we find the minimum value of $\sum_{i=1}^{99} C_{w_i}^3$.

Method 1. First, since the number of sets $(w_1, \ldots, w_{99})$ satisfying $w_1 + \cdots + w_{99} = 4752$ is finite, there must be a minimum. Second, intuitively, when $w_1 = \cdots = w_{99} = 48$, the sum $\sum_{i=1}^{99} C_{w_i}^3$ should be minimal. Assuming the opposite, then there must be an index i such that $w_i < 48$, and an index j such that $w_j > 48$. Change w_i to $w_i + 1$ and w_j to $w_j - 1$, we have a new set. Since

$$C_{w_i}^3 + C_{w_j}^3 - C_{w_i+1}^3 - C_{w_j-1}^3$$

$$= \frac{1}{2}(w_j - 1)(w_j - 2) - \frac{1}{2}w_i - (w_i - 1)$$

$$> 0 \text{ (since } w_j - 1 > w_i),$$

we know that the expression corresponding to the adjusted set is smaller than the original one, which is a contradiction. Thus the minimum value of $\sum_{i=1}^{99} C_{w_i}^3$ is $99C_{48}^3$. Thus the number of connected groups is no more than $C_{99}^4 - 99C_{48}^3 = 2052072$.

Method 2. From the power mean inequality, we have

$$\sum_{i=1}^{99} w_i^3 \geqslant \frac{1}{\sqrt{99}} \left(\sum_{i=1}^{99} w_i^2 \right)^{3/2},$$

$$\sum_{i=1}^{99} C_{w_i}^3 = \frac{1}{6} \sum_{i=1}^{99} w_i^3 - \frac{1}{2} \sum_{i=1}^{99} w_i^2 + \frac{1}{3} \sum_{i=1}^{99} w_i$$

$$\geqslant \frac{1}{6\sqrt{99}} \left(\sum_{i=1}^{99} w_i^2 \right)^{3/2} - \frac{1}{2} \sum_{i=1}^{99} w_i^2 + \frac{1}{3} \times 4752.$$

Notice that $\frac{1}{6\sqrt{99}} x^{3/2} - \frac{1}{2}x = \frac{1}{6}x\left(\sqrt{\frac{x}{99}} - 3\right)$ is strictly increasing, and

$$\sum_{i=1}^{99} w_i^2 \geqslant \frac{1}{99} \left(\sum_{i=1}^{99} w_i \right)^2 = 228096,$$

we have

$$\sum_{i=1}^{99} C_{w_i}^3 \geqslant \frac{1}{6\sqrt{99}} \times 228096^{3/2} - \frac{1}{2} \times 228096 + \frac{1}{3} \times 4752$$

$$= 1712304.$$

Therefore the number of connected groups is no more than

$$C_{99}^4 - 1712304 = 2052072.$$

Next we prove that equality can hold. First, set all the channels one-way, where the directions are set as follows: for $i < j$, if i, j has the same parity, then the channel linking A_i and A_j starts from A_i, otherwise it starts from A_j. Note that each point sends out exactly 49 one-way channels. Now change the 99 channels A_iA_{i+1} ($i = 1, 2, \ldots,$ 99, $A_{100} = A_1$) to two-way channels, then each point has exactly one outgoing channel and one incoming channel changed to two-way channels. Thus $w_i = l_i = 48$ ($i = 1, 2, \ldots, 99$). So the number of groups described above is $99C_{48}^3 = 1712304$.

Next we prove that each unconnected group must be one described above (i. e. , have a center, or a point with three outgoing channels). Actually, let (A_i, A_j, A_k, A_t) be an unconnected group without any center. If there is a two-way channel ij, then ik and jk cannot both start from k, or both end at k (if $j = i + 1$ or $j = i - 1$, then k will be larger or smaller than both i and j, and i, j have different parities; if $\{i, j\} = \{1, 99\}$, then i, j has the same parity, but k is larger than one and smaller than another). This holds for t as well, thus $ijkt$ is connected, which is a contradiction. Thus there is no two-way channel in the group. Since $ijkt$ has no two-way channels, there must be a station (say i) which has three incoming channels. Thus the values of j, k, t has only two cases: either smaller than i and has the same parity with i (called type I), or larger than i and has different parity with i (called type II). If j, k and t are all type I, then j, k and t has the same parity, thus the smallest one of them has three outgoing channels, which is a contradiction. If i, j and k are all type II, then the smallest one of i, j and t has three outgoing channels, which is a contradiction. If there are two of j, k and t (e. g. , j, k) being type I, then k, t are larger than i, j and have different parities, therefore one of k, t must have three outgoing channels, which is a contradiction. So it is proved that any unconnected group has a center. Thus, there are exactly 1712304 unconnected groups and hence 2052072 connected groups.

In conclusion, the maximum value is 2052072.

Example 5. Decompose 2006 as a sum of five positive integers x_1, x_2, x_3, x_4, x_5. Let $S = \sum_{1 \leqslant i < j \leqslant 5} x_i x_j$.

(1) What are the values of x_1, x_2, x_3, x_4, x_5 when S reaches its maximum?

(2) Furthermore, if for any $1 \leqslant i$, $j \leqslant 5$, we have $| x_i - x_j | \leqslant 2$, then what are the values of x_1, x_2, x_3, x_4, x_5 when S is minimal? Please explain your answer. (2006 Chinese National Mathematical Olympiad in Senior)

Solution. (1) First, the set of values of S is finite, so there must be a maximum and minimum value. We next prove that

If $x_1 + x_2 + x_3 + x_4 + x_5 = 2006$, and $S = \sum_{1 \leqslant i < j \leqslant 5} x_i x_j$ reaches its maximum, then

$$| x_i - x_j | \leqslant 1 (1 \leqslant i, j \leqslant 5). \qquad \text{①}$$

Assume that ① does not hold, let $x_1 - x_2 \geqslant 2$.

Set $x_1' = x_1 - 1$, $x_2' = x_2 + 1$, $x_i' = x_i (i = 2, 3, 4)$, we have

$$x_1' + x_2' = x_1 + x_2, \ x_1' \cdot x_2' = (x_1 - 1)(x_2 + 1)$$
$$= x_1 x_2 + x_1 - x_2 - 1 > x_1 x_2.$$

Rewrite S as

$$S = \sum_{1 \leqslant i < j \leqslant 5} x_i x_j$$
$$= x_1 x_2 + (x_1 + x_2)(x_3 + x_4 + x_5) + x_3 x_4 + x_3 x_5 + x_4 x_5,$$

and

$$S' = x_1' x_2' + (x_1' + x_2')(x_3 + x_4 + x_5) + x_3 x_4 + x_3 x_5 + x_4 x_5.$$

Then $S' - S = x_1' x_2' - x_1 x_2 > 0$, which contradicts the assumption that S is maximum at x_1, x_2, x_3, x_4, x_5. So we have $| x_i - x_j | \leqslant 1$ $(1 \leqslant i, j \leqslant 5)$. Thus when $x_1 = 402$, $x_2 = x_3 = x_4 = x_5 = 401$, S will be maximized.

(2) When $x_1 + x_2 + x_3 + x_4 + x_5 = 2006$ and $| x_i - x_j | \leqslant 2$ $(1 \leqslant i, j \leqslant 5)$, the only possibilities are as follows:

(I) 402, 402, 402, 400, 400;

(II) 402, 402, 401, 401, 400;

(III) 402, 401, 401, 401, 401.

The last two cases are obtained via adjusting case I : $x'_i = x_i - 1$, $x'_j = x_j + 1$. From the arguments in (1), each adjustment increases $S = \sum_{1 \leqslant i < j \leqslant 5} x_i x_j$. Thus S is minimized when $x_1 = x_2 = x_3 = 402$, $x_4 = x_5 = 400$.

Exercise 3

(1) Divide 1989 to 10 positive integers so that their product is maximized.

(2) In an increasing sequence of positive integers a_1, a_2, ... , a_m, ... , for any natural number m, let $b_m = \min\{n \mid a_n \geqslant m\}$. Given that $a_{19} = 85$, find the maximum value of $S = a_1 + a_2 + \cdots + a_{19} + b_1 + b_2 + \cdots + b_{85}$. (American Mathematics Olympiad in 1985)

(3) There are 155 birds on circle C. If the arc $P_i P_j \leqslant 10°$, then we call that the bird pair (P_i, P_j) is mutually visible. If it is allowed that more than one bird stays at the same position, find the smallest number of mutually visible bird pairs. (30th IMO shortlist)

(4) Given real numbers $P_1 \leqslant P_2 \leqslant P_3 \leqslant \cdots \leqslant P_n$, find real numbers $x_1 \geqslant x_2 \geqslant \cdots \geqslant x_n$, such that $d = (P_1 - x_1)^2 + (P_2 - x_2)^2 + \cdots + (P_n - x_n)^2$ is minimal.

(5) Given a set of points $P = \{p_1, p_2, \ldots, p_{1994}\}$ in the plane, no three being collinear. Divide the points in P to 83 groups so that each group has at least three points, and connect each pair of points in the same group with an edge while points in different groups are not connected. Now we have a graph G, let $m(G)$ be the number of triangles in the graph G.

(i) Find the minimum value of $m(G)$.

(ii) Let G' be the graph when $m(G)$ is minimal. Prove that: the points in G' can be 4-colored, so that G' has no triangle with the same color. (1994 Chinese National Mathematical Olympiad in Senior)

(6) 14 people are playing a Japanese chess round-robin tournament. Each one will play one game with each of the other 13 people, and the game has no draw. If there are three people such that each one of them has a win and a loss against the others, we call these three a "*triangle*". Find the maximum number of the triangles. (2002 Japanese National Mathematical Olympiad)

This approach is used to find extrema of symmetric polynomial functions. The basic idea of this approach is first to prove that the function has an extremum, then fix most of the variables and discuss the properties of the extremum point of the function with respect to the remaining variables. Due to the symmetry of the function, the extremum point of the function should have the same properties with respect to other variables. Consequently, the extremum point can be determined, and so is the extremum.

The basic steps are: prove that the extremum exists — fix as many variables as possible, turn the function into a single variable function or binary function and find the extremum point — find the extremum point of the function according to the symmetry — find the extremum.

Example 1. Let $0 < p \leqslant a$, b, c, d, $e \leqslant q$, find the maximum value of

$$F = (a + b + c + d + e)\left(\frac{1}{a} + \frac{1}{b} + \frac{1}{c} + \frac{1}{d} + \frac{1}{e}\right).$$

Solution. Since F is continuous in a closed domain, there must be a maximum value. Fix the variables a, b, c, d, let $a + b + c + d = A$, $\frac{1}{a} + \frac{1}{b} + \frac{1}{c} + \frac{1}{d} = B$. Then A and B are constants, and

$$F = (A + e)\left(B + \frac{1}{e}\right) = 1 + AB + eB + \frac{A}{e}.$$

Let $f(e) = eB + \frac{A}{e}$. It is easy to show that when $e \leqslant \sqrt{\frac{A}{B}}$, $f(e)$

is monotonically decreasing, and when $e \geqslant \sqrt{\dfrac{A}{B}}$, $f(e)$ is monotonically

increasing. Thus $\sqrt{\dfrac{A}{B}}$ is the minimum point of $f(e)$. Notice that $p \leqslant$ $e \leqslant q$, thus $f(e)$ can only reach its maximum at the endpoints $e = p$ or $e = q$. Due to the symmetry of the function, F can only reach its maximum when $a, b, c, d, e \in \{p, q\}$. Now, assume that there are k variables with value p and $5 - k$ variables with value q in $\{a, b, c, d, e\}$. Then we only need to maximize the following expression:

$$F(k) = [kp + (5 - k)q]\left(\frac{k}{p} + \frac{5 - k}{q}\right)$$

$$= k^2 + (5 - k)^2 + (5k - k^2)\left(\frac{p}{q} + \frac{q}{p}\right)$$

$$= \left[2 - \left(\frac{p}{q} + \frac{q}{p}\right)\right]k^2 + \left[5\left(\frac{p}{q} + \frac{q}{p}\right) - 10\right]k + 25.$$

In the above quadratic function $F(k)$, the coefficient of the quadratic term is $2 - \left(\dfrac{p}{q} + \dfrac{q}{p}\right) < 0$, and the x-coordinate of the peak of the corresponding parabola is $\dfrac{5}{2}$. Note that $k \in \mathbf{N}$, thus

$$F(k) < F(2) = F(3) = 13 + 6\left(\frac{p}{q} + \frac{q}{p}\right).$$

Also, when $a = b = c = p$, $c = d = q$, we have

$$F = 13 + 6\left(\frac{p}{q} + \frac{q}{p}\right).$$

So,

$$F_{\max} = 13 + 6\left(\frac{p}{q} + \frac{q}{p}\right) = 25 + 6\left(\sqrt{\frac{p}{q}} - \sqrt{\frac{q}{p}}\right)^2.$$

Example 2. Let $0 \leqslant x_i \leqslant 1$ $(1 \leqslant i \leqslant n)$, find the maximum value of $F = \sum_{1 \leqslant i < j \leqslant n} |x_i - x_j|$. (Sixth Canada Mathematical Olympiad)

Analysis and solution. The original solution to this question is very complicated, but using symmetry, we can obtain a much simpler solution.

First we prove the following lemma:

Let $0 \leqslant x \leqslant 1$. Then when $F(x) = |x - x_1| + |x - x_2| + \cdots + |x - x_n|$ reaches its maximum value, we must have $x \in \{0, 1\}$.

Actually, since $F(x)$ is continuous in the closed domain $0 \leqslant x \leqslant 1$, there must exist a maximum value of $F(x)$. Assume $x_1 \leqslant x_2 \leqslant \cdots \leqslant x_n$, and $x_0 = 0$, $x_{n+1} = 1$.

(1) If there is $i (i = 0, 1, 2, \ldots, n)$, such that $x_i < x < x_{i+1}$, then let $x' = x_i$ $\left(\text{when } i \leqslant \left[\dfrac{n}{2}\right]\right)$, or $x' = x_{i+1}\left(\text{when } i < \left[\dfrac{n}{2}\right]\right)$. We prove that $F(x) < F(x')$.

If $i \leqslant \left[\dfrac{n}{2}\right]$, let $d = x - x_i$. Adjust x to $x' = x_i$, then $|x - x_1|$, $|x - x_2|, \ldots, |x - x_i|$ will all decrease by d, and the total amount of decrease will be id. But $|x - x_{i+1}|$, $|x - x_{i+2}|, \ldots, |x - x_n|$ will all increase by d, which gives $(n - i)d$ in total. Thus after the adjustment, F has increased by $(n - 2i)d$, $F'(x) - F(x) = (n - 2i)d$. Since $i \leqslant \left[\dfrac{n}{2}\right] \leqslant \dfrac{n}{2}$, $F(x) \leqslant F(x')$, so x is not the maximum point, which is a contradiction.

If $i > \left[\dfrac{n}{2}\right]$, let $d = x_{i+1} - x$. Adjust x to $x' = x_{i+1}$, then $|x - x_1|$, $|x - x_2|, \ldots, |x - x_i|$ each increases by d, and the total amount of increase will be id. On the other hand, $|x - x_{i+1}|$, $|x - x_{i+2}|, \ldots,$ $|x - x_n|$ each decreases by d, which gives $(n - i)d$ in total. After the adjustment, F has increased by $(2i - n)d$, so $F'(x) - F(x) = (2i - n)d$. Since $i > \left[\dfrac{n}{2}\right]$ and i, n are both positive integers, we know that when n is even, $i > \dfrac{n}{2}$, and when n is odd, $i \geqslant \dfrac{n+1}{2} > \dfrac{n}{2}$. So $2i - n > 0$ always holds, thus $F'(x) > F(x)$. So x is not the maximum point, which is a contradiction.

(2) If there is i ($i = 1, 2, \ldots, n$), such that $x = x_i$, then let $x' = x_{i-1}$ $\left(\text{when } i \leqslant \left[\dfrac{n}{2}\right]\right)$, and $x' = x_{i+1}$ $\left(\text{when } i > \left[\dfrac{n}{2}\right]\right)$.

We prove that $F(x) < F(x')$.

If $i \leqslant \left[\dfrac{n}{2}\right]$, let $d = x - x_{i-1}$. We may adjust x to $x' = x_{i-1}$, so that $|x - x_1|$, $|x - x_2|$, $\ldots$, $|x - x_{i-1}|$ will all decrease by d, with a total of $(i-1)d$. But $|x - x_i|$, $|x - x_{i+1}|$, $\ldots$, $|x - x_n|$ will all increase by d, with a total of $(n - i + 1)d$. Thus after the adjustment, F has increased by $(n - 2i + 2)d$, thus $F'(x) - F(x) = (n - 2i + 2)d$. Since $i \leqslant \left[\dfrac{n}{2}\right] \leqslant \dfrac{n}{2}$, we know $F(x) < F(x')$, so x is not the point of maximum, which is a contradiction.

If $i > \left[\dfrac{n}{2}\right]$, let $d = x_{i+1} - x$. Adjust x to $x' = x_{i+1}$, then $|x - x_1|$, $|x - x_2|$, $\ldots$, $|x - x_i|$ will increase by d, with a total of id. Also $|x - x_{i+1}|$, $|x - x_{i+2}|$, $\ldots$, $|x - x_n|$ will decrease by d, with a total of $(n - i)d$. After the adjustment, F increases by $(2i - n)d$, so $F'(x) - F(x) = (2i - n)d$. Since $i > \left[\dfrac{n}{2}\right]$, and i, n are positive integers, we know that when n is even, $i > \dfrac{n}{2}$, and when n is odd, $i \geqslant \dfrac{n+1}{2} > \dfrac{n}{2}$. So we always have $2i - n > 0$, $F'(x) > F(x)$, so x is not the maximum point, which is a contradiction.

In conclusion, we must have $x \in \{0, 1\}$ when F reaches its maximum.

Now we solve the original problem.

Since F is continuous in a closed domain, F must have a maximum value. Fix x_2, x_3, $\ldots$, x_n. Then $F(x_1)$ is a function of x_1:

$$F(x_1) = |x_1 - x_2| + |x_1 - x_3| + \cdots + |x_1 - x_n| + \sum_{2 \leqslant i < j \leqslant n} |x_i - x_j|.$$

Now, $F(x_1)$ reaches its maximum if and only if $|x_1 - x_2| + |x_1 - x_3| + \cdots + |x_1 - x_n|$ reaches its maximum. Since $0 \leqslant x_i \leqslant 1$, from the above lemma, when $F(x_1)$ reaches its maximum, we have $x_1 \in \{0, 1\}$. By symmetry, $x_i \in \{0, 1\} (1 \leqslant i \leqslant n)$.

Now, assume there are k 0 s and $n - k$ 1s in x_i when F reaches its maximum. Then

$$F \leqslant (0 - 0) \times C_k^2 + (1 - 1) \times C_{n-k}^2 + C_k^1 C_{n-k}^1$$

$$= k(n - k) \leqslant \left[\frac{k + (n - k)}{2}\right]^2 = \frac{n^2}{4}.$$

Since F is an integer, then $F \leqslant \left[\dfrac{n^2}{4}\right]$.

The equality holds when $x_1 = x_2 = \cdots = x_{\left[\frac{n}{2}\right]} = 0$, $x_{\left[\frac{n}{2}\right]+1} = \cdots = x_n = 1$. So the maximum value of F is $\left[\dfrac{n^2}{4}\right]$.

Example 3. Assume that real numbers x_1, x_2, ..., x_{1997} satisfy the following two conditions:

(1) $-\dfrac{1}{\sqrt{3}} \leqslant x_i \leqslant \sqrt{3} (i = 1, 2, \ldots, 1997)$;

(2) $x_1 + x_2 + \cdots + x_{1997} = -318\sqrt{3}$.

Find the maximum value of $x_1^{12} + x_2^{12} + \cdots + x_{1997}^{12}$, and justify your answer. (1997 China Mathematical Olympiad)

Solution. Since $x_1 + x_2 + \cdots + x_{1997}$ is continuous in a closed domain, we know that $x_1^{12} + x_2^{12} + \cdots + x_{1997}^{12}$ has a maximum value. Fix x_3, x_4, ..., x_{1997}. Then $x_1 + x_2 = c$ (constant). Let $x_1 = x$. Then $x_2 = c - x$. Moreover

$$x_1^{12} + x_2^{12} + \cdots + x_{1997}^{12} = x^{12} + (x - c)^{12} + A = f(x).$$

Since $f''(x) = 132x^{10} + 132(x - c)^{10} > 0$, $f(x)$ is a convex function, so when $f(x)\left(-\dfrac{1}{\sqrt{3}} \leqslant x \leqslant \sqrt{3}\right)$ reaches its maximum, we have $x \in \left\{-\dfrac{1}{\sqrt{3}}, \sqrt{3}\right\}$. By symmetry, for any two variables x_i, $x_j (1 \leqslant i < j \leqslant 1997)$, when $x_1^{12} + x_2^{12} + \cdots + x_{1997}^{12}$ reaches its maximum, one of x_i, x_j should be in $\left\{-\dfrac{1}{\sqrt{3}}, \sqrt{3}\right\}$. Thus, when $x_1^{12} + x_2^{12} + \cdots + x_{1997}^{12}$ reaches its maximum, at least 1996 numbers in x_1, x_2, ..., x_{1997} are in $\left\{-\dfrac{1}{\sqrt{3}}, \sqrt{3}\right\}$. Assume that in x_1, x_2, ..., x_{1997} there are u numbers equaling $-\dfrac{1}{\sqrt{3}}$, v numbers equaling $\sqrt{3}$, and w numbers in $\left(-\dfrac{1}{\sqrt{3}}, \sqrt{3}\right)$, where $w = 0$ or 1. When $w = 1$, we denote the variable in $\left(-\dfrac{1}{\sqrt{3}}, \sqrt{3}\right)$ by t. Then $u + v + w = 1997$, $-\dfrac{1}{\sqrt{3}}u + \sqrt{3}v + tw = -318\sqrt{3}$.

Eliminating u, we have $4v + (\sqrt{3}t + 1)w = 1043$. Thus $(\sqrt{3}t + 1)w$ is an integer, and $(\sqrt{3}t + 1)w \equiv 1043 \equiv 3 \pmod{4}$. Since $-\dfrac{1}{\sqrt{3}} < t < \sqrt{3}$, we have $0 < \sqrt{3}t + 1 < 4$; since $w = 0$ or 1, we have $0 \leqslant (\sqrt{3}t + 1)w < 4$; so $(\sqrt{3}t + 1)w = 3$, thus $w \neq 0$, so $w = 1$, $t = \dfrac{2}{\sqrt{3}}$ and $4v + 3 = 1043$, thus $v = 260$, so $u = 1997 - v - w = 1736$.

In conclusion, the maximum value of $x_1^{12} + x_2^{12} + \cdots + x_{1997}^{12}$ is

$$\left(-\frac{1}{\sqrt{3}}\right)^{12} u + (\sqrt{3})^{12} v + t^{12} = \left(\frac{1}{3}\right)^6 \times 1736 + 3^6 \times 260 + \left(\frac{2}{\sqrt{3}}\right)^{12}$$

$$= 189548.$$

Example 4. Let x_1, x_2, ..., x_n be in an interval of length 1. Define $x = \dfrac{1}{n}\sum_{j=1}^{n} x_j$, $y = \dfrac{1}{n}\sum_{j=1}^{n} x_j^2$. Find the maximum value of $f = y - x^2$.

(National Mathematical Olympiad with Senior)

Solution. Let x_1, x_2, ..., $x_n \in [a, a+1]$ $(a \in \mathbf{R})$. When $n = 1$, $f = 0$, thus $f_{\max} = 0$.

When $n > 1$, fix x_2, x_3, ..., x_n; then

$$f = y - x^2 = \frac{1}{n}\sum_{j=1}^{n} x_j^2 - \left(\frac{1}{n}\sum_{j=1}^{n} x_j\right)^2$$

$$= \frac{n-1}{n^2}x_1^2 - \left(\frac{2}{n^2}\sum_{j=2}^{n} x_j\right)x_1 + \frac{1}{n}\sum_{j=2}^{n} x_j^2 - \left(\frac{1}{n}\sum_{j=2}^{n} x_j\right)^2.$$

Since f is a quadratic function with respect to x_1 in $[a, a+1]$, and the coefficient of the quadratic term is $\dfrac{n-1}{n^2} > 0$, we have

$$f \leqslant \max\{f(a, x_2, x_3, \ldots, x_n), f(a+1, x_2, x_3, \ldots, x_n)\}.$$

This means that, when f reaches its maximum, $x_1 \in \{a, a+1\}$. By symmetry, when f reaches its maximum, $x_1, x_2, \ldots, x_n \in \{a, a+1\}$; thus

$$f \leqslant \max_{x_i \in \{a, a+1\}, 1 \leqslant i \leqslant n} \{f(x_1, x_2, x_3, \ldots, x_n)\}.$$

Assume when f reaches its maximum that there are s variables equaling a and $n-s$ variables equaling $a+1$, then

$$\max_{x_i \in \{a, a+1\}, 1 \leqslant i \leqslant n} \{f(x_1, x_2, x_3, \ldots, x_n)\}$$

$$= \frac{1}{n}[sa^2 + (n-s)(a+1)^2] - \frac{1}{n^2}[sa + (n-s)(a+1)]^2$$

$$= \frac{1}{n^2}s(n-s) \leqslant \begin{cases} \dfrac{n^2-1}{4n^2} & \text{if } n \text{ is odd,} \\[2mm] \dfrac{1}{4} & \text{if } n \text{ is even.} \end{cases}$$

When there are $\left[\dfrac{n+1}{2}\right]$ variables equaling a and $n - \left[\dfrac{n+1}{2}\right]$ variables equaling $a+1$, then the following equality holds:

$$f_{\max} = \begin{cases} \dfrac{n^2-1}{4n^2} & \text{if } n \text{ is odd,} \\[2mm] \dfrac{1}{4} & \text{if } n \text{ is even.} \end{cases}$$

Exercise 4

(1) Let $0 < p \leqslant a_1, a_2, \ldots, a_n \leqslant q$, find the maximum value of

$$F = (a_1 + a_2 + \cdots + a_n)\left(\frac{1}{a_1} + \frac{1}{a_2} + \cdots + \frac{1}{a_n}\right).$$

(2) Let $x_i \in \mathbf{R}$, $|x_i| \leqslant 1 (1 \leqslant i \leqslant n)$, find the minimum value of $F = \sum_{1 \leqslant i < j \leqslant n} x_i x_j$. (USSR Mathematical Olympiad)

(3) Given natural number $n > 2$, and a constant λ, find the maximum and minimum values of function $F = x_1^2 + x_2^2 + \cdots + x_n^2 + \lambda x_1 x_2 \cdots x_n$. Here $x_1, x_2, \ldots, x_n$ are non-negative real numbers such that $x_1 + x_2 + \cdots + x_n = 1$. (1994 China National Team Test)

Chapter 5   Polishing Transform

The extremum of some functions may not be easy to prove to exist, but it may be easily guessed according to the heuristics of the problem. In this case, we can apply a transform to: first, adjust some variable to the extreme point, and (for example when the sum of all variables is fixed) transform the difference between the variable and the corresponding variable of the extreme point to other variables; then we justify that the transform keeps the function value nonincreasing or nondecreasing. We then apply such transform repeatedly, until all of the variables are adjusted to match the extremum point. In this way, we can obtain the extremum of the function. Such a transform is called a polishing transform, and the inequality used to establish this is called the polishing tool.

Usually, polishing transform takes one of the following forms:

(1) For the *"pairing"* type of extrema, such as when the variable with minimum value is paired with the variable with maximum value, we can adjust the paired variables to the extreme point, and then adjust the other pairs.

(2) For the *"uniform"* type of extrema, such as when the variables have identical values, we can first adjust the variable with the minimum value to match the extremum point, and change the variables with larger values correspondingly.

(3) For the *"concentration"* type of extrema, such as when one variable takes a maximum value and the other variables take small values, we can first adjust the variable with the minimal value to match the extreme point, and then adjust the rest variables to match the extreme point.

Essentially, polishing transform is nothing but enlarging and shrinking, just that the way to enlarge or to shrink is applying polishing transform.

Example 1. Given $2n$ real numbers $a_1 \leqslant a_2 \leqslant \cdots \leqslant a_n$, $b_1 \leqslant b_2 \leqslant \cdots \leqslant b_n$, let $F = a_1 b_{i_1} + a_2 b_{i_2} + \cdots + a_n b_{i_n}$, where the sequence $b_{i_1}, b_{i_2}, \ldots, b_{i_n}$ is a permutation of $b_1, b_2, \ldots, b_n$. Find the maximum and minimum values of F. (The rearrangement inequality)

Analysis. Intuitively, we may guess that when the minimum one of a_i is matched with the minimum one of b_i, and the maximum one of a_i is matched with the maximum one of b_i, we get the maximum value of F. Thus we may apply polishing transform to obtain $a_1 b_1, a_2 b_2, \ldots, a_n b_n$.

Solution. If the expression of F does not contain $a_1 b_1$, then consider the two terms containing a_1, b_1 respectively, $a_1 b_{i_1}$ and $a_j b_1$, and adjust them to $a_1 b_1$, $a_j b_{i_1}$. Note that

$$a_1 b_1 + a_j b_{i_1} - (a_1 b_{i_1} + a_j b_1) = (a_1 - a_j)(b_1 - b_{i_1})$$
$$\geqslant 0 \text{ (polishing tool)}.$$

Thus we can see that, after the adjustment, the value of F does not decrease.

Fixing the product $a_1 b_1$ and consider the remaining $n - 1$ products, and continue this process until we obtain $a_1 b_1, a_2 b_2, \ldots, a_n b_n$, then the value of F does not decrease. Thus $F_1 = a_1 b_1 + a_2 b_2 + \cdots + a_n b_n$ is the maximum value of F.

Similarly, when the terms $a_1 b_n, a_2 b_{n-1}, \ldots, a_n b_1$ appear, F reaches its minimum.

Example 2. Assume that A, B, C are the three angles of a triangle. Find the maximum value of $\sin A + \sin B + \sin C$.

Analysis and Solution. It is a simple problem which can be solved in multiple ways, and the polishing transform is a relatively complicated one. However, the process of finding the polishing tools is very typical.

First, we guess that the extremum point is $\left(\dfrac{\pi}{3}, \dfrac{\pi}{3}, \dfrac{\pi}{3}\right)$.

Second, it is easy to find the polishing method: $(A, B, C) \rightarrow \left(\dfrac{\pi}{3}, A + B - \dfrac{\pi}{3}, C\right)$; that is, letting $A' = \dfrac{\pi}{3}$, $B' = A + B - \dfrac{\pi}{3}$, $C' = C$, we hope that after polishing, the function value increases. That is,

$$\sin A' + \sin B' + \sin C' = 2\sin\frac{A' + B'}{2}\cos\frac{A' - B'}{2} + \sin C$$

$$= 2\sin\frac{A + B}{2}\cos\frac{A' - B'}{2} + \sin C$$

$$\geqslant 2\sin\frac{A + B}{2}\cos\frac{A - B}{2} + \sin C$$

$$= \sin A + \sin B + \sin C. \qquad ①$$

To make the equation ① hold, we need that $\dfrac{\mid A' - B' \mid}{2} < \dfrac{\mid A - B \mid}{2}$, and a sufficient condition for this is that A is the minimum of A, B, C, and B is the maximum one of A, B, C. This lead to the following lemma (polishing tool): If $0° < A \leqslant 60° \leqslant B < 180°$, then

$$\sin A + \sin B \leqslant \sin 60° + \sin (A + B - 60°).$$

Actually,

$$A + B - 120° = A + B - 2 \times 60° \geqslant A + B - 2B = A - B;$$
$$A + B - 120° = A + B - 2 \times 60° \leqslant A + B - 2A = B - A.$$

So $\mid A + B - 120° \mid \leqslant B - A$.

Thus

$$\sin A + \sin B = 2\sin\frac{A + B}{2}\cos\frac{A - B}{2}$$

$$\leqslant 2\sin\frac{A + B}{2}\cos\frac{A + B - 120°}{2}$$

$$= \sin 60° + \sin(A + B - 60°).$$

The lemma is proved.

We come back to the original problem. Assume $A \leqslant B \leqslant C$, then $A \leqslant 60° \leqslant C$, from the lemma, we have

$$\sin A + \sin B + \sin C \leqslant \sin 60° + \sin(A + C - 60°) + \sin B.$$

Since $(A + C - 60°) + B = A + B + C - 60° = 120°$, we have $0° < A + C - 60° \leqslant 60° \leqslant B < 180°$. Using the lemma again, we have

$$\sin A + \sin B + \sin C \leqslant \sin 60° + \sin(A + C - 60°) + \sin B$$

$$\leqslant \sin 60° + \sin 60° + \sin 60° = 3\sin 60° = \frac{3\sqrt{3}}{2}.$$

Thus the maximum value of $\sin A + \sin B + \sin C$ is $\dfrac{3\sqrt{3}}{2}$.

Example 3. Let $x_i \geqslant 0 (1 \leqslant i \leqslant n)$, $\sum_{i=1}^{n} x_i = 1$, $n \geqslant 2$. Find the maximum value of $F = \sum_{1 \leqslant i < j \leqslant n} x_i x_j (x_i + x_j)$.

Analysis and Solution. Since F is continuous in a closed domain, the maximum value of F exists. We may try to use partial adjustments. If we fix x_2, x_3, ..., x_n, then x_1 is also fixed (so we cannot make adjustments). Thus we fix $n - 2$ numbers. Assume x_3, x_4, ..., x_n are fixed, then

$$F = x_1 x_2 (x_1 + x_2) + \cdots + x_1 x_n (x_1 + x_n) + x_2 x_3 (x_2 + x_3)$$

$$+ \cdots + x_2 x_n (x_2 + x_n) + \sum_{3 \leqslant i < j \leqslant n} x_i x_j (x_i + x_j),$$

where $x_1 + x_2 = 1 - (x_2 + x_3 + \cdots + x_n) = p$ (constant).

Note that $x_2 = p - x_1$, we know that F is a quadratic function of x_1: $f(x_1) = Ax_1^2 + Bx_1 + C (0 \leqslant x_1 \leqslant p)$. But the ranges of the coefficients and the variables of the function are complicated, making the expression of the extremum point quite complicated. So we need another way to find the extremum.

First we try to find the extremum point. To do this, we first consider the simple cases.

When $n = 2$, $F = x_1 x_2 (x_1 + x_2) = x_1 x_2 \leqslant \dfrac{(x_1 + x_2)^2}{4} = \dfrac{1}{4}$.

However, $n = 2$ is too special to be typical. So we try again.

When $n = 3$, $F = x_1 x_2 (x_1 + x_2) + x_1 x_3 (x_1 + x_3) + x_2 x_3 (x_2 + x_3)$. Fix x_3, then

$$\begin{aligned}
F &= x_1 x_2 (x_1 + x_2) + (x_1^2 + x_2^2) x_3 + (x_1 + x_2) x_3^2 \\
&= x_1 x_2 (1 - x_3) + [(x_1 + x_2)^2 - 2 x_1 x_2] x_3 + (1 - x_3) x_3^2 \\
&= x_1 x_2 (1 - 3 x_3) + (1 - x_3)^2 x_3 + (1 - x_3) x_3^2.
\end{aligned}$$

Now, to use the inequality $x_1 x_2 \leqslant \dfrac{(x_1 + x_2)^2}{4}$ to enlarge $x_1 x_2 (1 -$

$3 x_3)$, we need that $1 - 3 x_3 \geqslant 0$. Thus, we make the assumption $x_1 \geqslant x_2 \geqslant x_3$. Then

$$\begin{aligned}
F &= x_1 x_2 (1 - 3 x_3) + (1 - x_3)^2 x_3 + (1 - x_3) x_3^2 \\
&\leqslant (1 - 3 x_3) \frac{(x_1 + x_2)^2}{4} + (1 - x_3)^2 x_3 + (1 - x_3) x_3^2 \\
&= (1 - 3 x_3) \frac{(1 - x_3)^2}{4} + (1 - x_3)^2 x_3 + (1 - x_3) x_3^2 \\
&\leqslant \frac{1}{4} \text{ (by differentiation)}.
\end{aligned}$$

Equality holds when $x_1 = x_2 = \dfrac{1}{2}$, $x_3 = 0$.

Therefore we may guess that, in general, the extremum point is $\left(\dfrac{1}{2}, \dfrac{1}{2}, 0, 0, \ldots, 0 \right)$.

Apply polishing transform: when $n \geqslant 3$, for the variables $(x_1, x_2, \ldots, x_n)$, assume that $x_1 \geqslant x_2 \geqslant \cdots \geqslant x_n$. Let $x_1' = x_1$, $x_2' = x_2, \ldots$, $x_{n-2}' = x_{n-2}$, $x_{n-1}' = x_{n-1} + x_n$, $x_n' = 0$, we get a new set of variables $(x_1, x_2, \ldots, x_{n-2}, x_{n-1} + x_n, 0)$, and the corresponding value of F is

$$\begin{aligned}
F' &= \sum_{1 \leqslant i < j \leqslant n-2} x_i x_j (x_i + x_j) \text{ (denote this by } A) \\
&\quad + \sum_{i=1}^{n-2} x_{n-1}' x_i (x_{n-1}' + x_i) + \sum_{i=1}^{n-1} x_n' x_i (x_n' + x_i) \\
&= A + \sum_{i=1}^{n-2} (x_{n-1} + x_n) x_i (x_{n-1} + x_n + x_i)
\end{aligned}$$

$$= A + \sum_{i=1}^{n-2}(x_{n-1}^2 + 2x_{n-1}x_n + x_{n-1}x_i + x_n^2 + x_n x_i)x_i$$

$$= A + \sum_{i=1}^{n-2}x_{n-1}x_i(x_{n-1}+x_i) + \sum_{i=1}^{n-2}x_n x_i(x_n+x_i) + 2x_{n-1}x_n\sum_{i=1}^{n-2}x_i$$

$$= A + \sum_{i=1}^{n-2}x_{n-1}x_i(x_{n-1}+x_i) + \sum_{i=1}^{n-1}x_n x_i(x_n+x_i) - x_{n-1}x_n(x_{n-1}+x_n)$$

$$+ 2x_{n-1}x_n\sum_{i=1}^{n-2}x_i$$

$$= F + x_{n-1}x_n\left(2\sum_{i=1}^{n-2}x_i - x_{n-1} - x_n\right)$$

$$= F + [2 - 3(x_{n-1}+x_n)]\,x_{n-1}x_n.$$

Since $\sum_{i=1}^n x_i = 1$ and $x_1 \geqslant x_2 \geqslant \cdots \geqslant x_n$, we have

$$\frac{x_{n-1}+x_n}{2} \leqslant \frac{x_1+x_2+\cdots+x_n}{n} = \frac{1}{n}.$$

So $x_{n-1}+x_n \leqslant \dfrac{2}{n}$ and $n \geqslant 3$, thus $x_{n-1}+x_n \leqslant \dfrac{2}{n} \leqslant \dfrac{2}{3}$, therefore $2 - 3(x_{n-1}+x_n) \geqslant 0$, so $F' \geqslant F$.

As long as the number of nonzero variables is at least 3, the above transform can be continued; after at most $n-2$ steps, $n-2$ of the variables can be changed to 0, and it reduces to the case $n=2$, thus F reaches its maximum value $\dfrac{1}{4}$ at the maximum point $\left(\dfrac{1}{2}, \dfrac{1}{2}, 0, 0, \ldots, 0\right)$.

Example 4. Assume that all of the coefficients of $f(x) = ax^2 + bx + c$ are positive, and $a+b+c=1$. For all positive numbers $x_1, x_2, \ldots, x_n$, satisfying that $x_1 x_2 \ldots x_n = 1$, find the minimum value of $f(x_1)f(x_2)\ldots f(x_n)$. (USSR Mathematical Olympiad)

Solution. $f(1) = a+b+c = 1$.

If $x_1 = x_2 = \cdots = x_n = 1$, then $f(x_1)f(x_2)\ldots f(x_n) = 1$.

If not all of $x_1, x_2, \ldots, x_n$ are 1, since $x_1 x_2 \ldots x_n = 1$, one of them must be smaller than 1, and one of them larger than 1. Assume

$x_1 > 1$, $x_2 < 1$, replace x_1, x_2 by 1, $x_1 x_2$, and compute:

$$f(x_1)f(x_2) = (ax_1^2 + bx_1 + c)(ax_2^2 + bx_2 + c)$$
$$= a^2 x_1^2 x_2^2 + b^2 x_1 x_2 + c^2 + ab(x_1^2 x_2 + x_1 x_2^2)$$
$$+ ac(x_1^2 + x_2^2) + bc(x_1 + x_2),$$
$$f(1)f(x_1 x_2) = (a + b + c)(ax_1^2 x_2^2 + bx_1 x_2 + c)$$
$$= a^2 x_1^2 x_2^2 + b^2 x_1 x_2 + c^2 + ab(x_1^2 x_2^2 + x_1 x_2)$$
$$+ ac(x_1^2 x_2^2 + 1) + bc(x_1 x_2 + 1),$$

$$f(x_1)f(x_2) - f(1)f(x_1 x_2)$$
$$= abx_1 x_2(x_1 + x_2 - x_1 x_2 - 1) + ac(x_1^2 + x_2^2 - x_1 x_2 x_2^2 - 1)$$
$$+ bc(x_1 + x_2 - x_1 x_2 - 1)$$
$$= - abx_1 x_2(x_1 - 1)(x_2 - 1) - ac(x_1^2 - 1)(x_2^2 - 1)$$
$$- bc(x_1 - 1)(x_2 - 1) > 0.$$

Repeating the above polishing transform, we get that

$$f(x_1)f(x_2)\ldots f(x_n) \geqslant f(1)f(1)\ldots f(1) = 1.$$

So $f(x_1)f(x_2)\ldots f(x_n)$ has a minimum value of 1.

Example 5. For nonnegative real numbers x_1, x_2, $\ldots$, x_n, which satisfy $x_1 + x_2 + \cdots + x_n = 1$, find the maximum value of $S = \sum_{j=1}^{n}(x_j^4 - x_j^5)$. (40th IMO China National Team Selection)

Analysis. Considering the cases $n = 2$, 3, we can find that when $\sum_{j=1}^{n}(x_j^4 - x_j^5)$ reaches its maximum value, there can be at most 2 of x_1, x_2, $\ldots$, x_n that are nonzero. Consider a " *polishing transformation*" $(x, y) \to (x + y, 0)$, we hope that $(x + y)^4 - (x + y)^5 + 0^4 - 0^5 > x^4 - x^5 + y^4 - y^5$. This inequality is equivalent to $4x^2 + 4y^2 + 6xy > 5x^3 + 5y^3 + 10x^2 y + 10xy^2$. The left-hand side of the inequality is

$$\frac{7}{2}(x^2 + y^2) + \frac{1}{2}(x^2 + y^2) + 6xy \geqslant \frac{7}{2}(x^2 + y^2) + xy + 6xy$$

$$= \frac{7}{2}(x + y)^2,$$

and the right-hand side does not exceed $5x^3 + 5y^3 + 15x^2y + 15xy^2 = 5(x + y)^3$. Thus, a sufficient condition to make the inequality hold is that $\frac{7}{2}(x + y)^2 > 5(x + y)^3$, or $x + y < \frac{7}{10}$. Thus, we get the following lemma.

Lemma. If $x + y < \frac{7}{10}$, then $(x + y)^4 - (x + y)^5 > x^4 - x^5 + y^4 - y^5$.

Solution. Assume that the number of nonzero numbers in x_1, x_2, ... , x_n is k, we may assume $x_1 \geqslant x_2 \geqslant \cdots \geqslant x_k > 0$, $x_{k+1} = x_{k+2} = \cdots = x_n = 0$. If $k \geqslant 3$, let $x_i' = x_i$ $(i = 1, 2, \ldots, k - 2)$, $x_{k-1}' = x_{k-1} + x_k$, $x_k' = x_{k+1}' = \cdots = x_n' = 0$. Since $x_{k-1} + x_k \leqslant \frac{2}{n} \leqslant \frac{2}{3} < \frac{7}{10}$, from the lemma, we have

$$\sum_{j=1}^{n}(x'^4_j - x'^5_j) > \sum_{j=1}^{n}(x^4_j - x^5_j).$$

As long as the number of nonzero numbers is no less than 3, the above adjustment can be carried on. After at most $n - 2$ steps, x_3, ... , x_n can be adjusted to 0, while S does not decrease. Now denote (x_1, x_2) as (a, b), then $a + b = 1$, and

$$\begin{aligned}
S &= a^4(1 - a) + b^4(1 - b) = a^4b + ab^4 = ab(a^3 + b^3) \\
&= ab(a + b)(a^2 - ab + b^2) = ab[(a + b)^2 - 3ab] \\
&= ab(1 - 3ab) = \frac{1}{3}(3ab)(1 - 3ab) \leqslant \frac{1}{3} \times \frac{1}{4} = \frac{1}{12}.
\end{aligned}$$

Since when $x_1 = \frac{3 + \sqrt{3}}{6}$, $x_2 = \frac{3 - \sqrt{3}}{6}$, $x_3 = \cdots = x_n = 0$, we have $S = \frac{1}{12}$, and we know that $S_{\max} = \frac{1}{12}$.

Example 6. Let x, y, z be nonnegative real numbers, which satisfy $x + y + z = 1$. Find the minimum value of $Q = \sqrt{2 - x} + \sqrt{2 - y} + \sqrt{2 - z}$.

Solution. By symmetry, assume $x \leqslant y \leqslant z$. Let $x' = 0$, $y' = y$, $z' = z + x - x'$, then $x' \geqslant 0$, $y' \geqslant 0$, $z' \geqslant 0$, and $z' - z = x - x'$, $y' + z' = x + y + z - x' = x + y + z = 1$, then

$$Q - (\sqrt{2-x'} + \sqrt{2-y'} + \sqrt{2-z'})$$

$$= \sqrt{2-x} + \sqrt{2-y} + \sqrt{2-z} - (\sqrt{2-x'} + \sqrt{2-y'} + \sqrt{2-z'})$$

$$= (\sqrt{2-x} - \sqrt{2-x'}) + (\sqrt{2-z} - \sqrt{2-z'})$$

$$= \frac{x'-x}{\sqrt{2-x} + \sqrt{2-x'}} + \frac{z'-z}{\sqrt{2-z} + \sqrt{2-z'}}$$

$$= (x-x') \left[\frac{-1}{\sqrt{2-x} + \sqrt{2-x'}} + \frac{1}{\sqrt{2-z} + \sqrt{2-z'}} \right]$$

$$= x \cdot \frac{(\sqrt{2-x} + \sqrt{2-x'}) - (\sqrt{2-z} + \sqrt{2-z'})}{(\sqrt{2-x} + \sqrt{2-x'})(\sqrt{2-z} + \sqrt{2-z'})}$$

$$= x \cdot \frac{(\sqrt{2-x} - \sqrt{2-z}) + (\sqrt{2-x'} - \sqrt{2-z'})}{(\sqrt{2-x} + \sqrt{2-x'})(\sqrt{2-z} + \sqrt{2-z'})}$$

$$= x \cdot \frac{\dfrac{z-x}{\sqrt{2-x} + \sqrt{2-z}} + \dfrac{z'-x'}{\sqrt{2-x'} - \sqrt{2-z'}}}{(\sqrt{2-x} + \sqrt{2-x'})(\sqrt{2-z} + \sqrt{2-z'})}$$

$$= x \cdot \frac{\dfrac{z-x}{\sqrt{2-x} + \sqrt{2-z}} + \dfrac{z'}{\sqrt{2-x'} - \sqrt{2-z'}}}{(\sqrt{2-x} + \sqrt{2-x'})(\sqrt{2-z} + \sqrt{2-z'})} \geqslant 0,$$

So,

$$Q \geqslant \sqrt{2-x'} + \sqrt{2-y'} + \sqrt{2-z'}$$

$$= \sqrt{2} + \sqrt{2-y'} + \sqrt{2-z'}. \tag{①}$$

Since $(\sqrt{2-y'} + \sqrt{2-z'})^2 = (2-y') + (2-z') + 2\sqrt{2-y'}\sqrt{2-z'}$

$$= 4 - (y' + z')$$

$$+ 2\sqrt{4 - 2(y' + z') + y'z'}$$

$$= 4 - 1 + 2\sqrt{4 - 2 \cdot 1 + y'z'}$$

$$= 3 + 2\sqrt{2} + y'z'$$
$$\geqslant 3 + 2\sqrt{2} = (1 + \sqrt{2})^2,$$

we have

$$\sqrt{2 - y'} + \sqrt{2 - z'} \geqslant 1 + \sqrt{2}. \qquad ②$$

From ①②, $Q \geqslant 1 + 2\sqrt{2}$. When $x = y = 0$, $z = 1$, $Q = 1 + 2\sqrt{2}$, So the minimum value of Q is $1 + 2\sqrt{2}$.

Exercise 5

(1) Let $x_i \geqslant 0 (1 \leqslant i \leqslant n)$, $\sum_{i=1}^{n} x_i \leqslant \frac{1}{2}$, $n \geqslant 2$. Find the minimum value of $F = (1 - x_1)(1 - x_2)\cdots(1 - x_n)$.

(2) Let $x_i \geqslant 0 (1 \leqslant i \leqslant n, n \geqslant 4)$, $\sum_{i=1}^{n} x_i = 1$. Find the maximum value of $F = \sum_{i=1}^{n} x_i x_{i+1}$.

(3) Assume $x_i \geqslant 0 (1 \leqslant i \leqslant n)$, $\sum_{i=1}^{n} x_i = \pi$, $n \geqslant 2$. Find the maximum value of $F = \sum_{i=1}^{n} \sin^2 x_i$. (26th IMO alternative)

(4) Let $0 < a_1 \leqslant a_2 \leqslant \cdots \leqslant a_n < \pi$, $a_1 + a_2 + \cdots + a_n = A$. Find the maximum value of $\sin a_1 + \sin a_2 + \cdots + \sin a_n$.

(5) Let x_1, x_2, x_3, x_4 be positive real numbers, and $x_1 + x_2 + x_3 + x_4 = \pi$. Find the maximum value of

$$A = \left(2\sin^2 x_1 + \frac{1}{\sin^2 x_1}\right)\left(2\sin^2 x_2 + \frac{1}{\sin^2 x_2}\right)\left(2\sin^2 x_3 + \frac{1}{\sin^2 x_3}\right)\left(2\sin^2 x_4 + \frac{1}{\sin^2 x_4}\right).$$

Chapter 6　　　　　Space Estimates

Consider the following problem. Let X be a given set and A be a subset of X which has some property, find the maximum value of $|A|$. To solve it, we can sort the elements of the set X in some way, and estimate the distance between two consecutive elements in A, thus obtaining an estimate for the number of elements in A. This approach is called space estimation.

Example 1. Let $M = \{1, 2, \ldots, 2005\}$, and A be a subset of M. If for any a_i, $a_j \in A$ with $a_i \neq a_j$, exactly one isosceles triangle can be determined with a_i, a_j as its side lengths, find the maximum value of $|A|$.

Analysis. First we consider under what situation can two numbers a, $b(a < b)$ determine a unique isosceles triangle. Since (a, b, b) form an isosceles triangle, we know that (a, a, b) cannot form an isosceles triangle. That is, $2a \leqslant b$. Therefore, we can apply the method of space estimation.

Solution. When $a < b$, a, b, b form an isosceles triangle. Thus a, b determine only one isosceles triangle if and only if a, a, b cannot form an isosceles triangle, which is equivalent to $2a \leqslant b$. Thus

$$2005 \geqslant a_n \geqslant 2a_{n-1} \geqslant \cdots \geqslant 2^{n-1}a_1 \geqslant 2^{n-1},$$

thus $n \leqslant 11$.

Next, let $A = \{1, 2, 4, \ldots, 1024\}$, then $|A| = 11$. For any a_i, $a_j \in A$, let $a_i = 2^i$, $a_j = 2^j$, we have $2a_i = 2^{i+1} \leqslant 2^j = a^j$, thus a_i, a_j can form only one isosceles triangle (a_i, a_j, a_j). Therefore $|A|$ satisfies the requirement.

In conclusion, the maximum value of $|A|$ is 11.

Example 2. Let A be a subset of the set $\mathbf{N}^*$ of positive integers. For any x, $y \in A$, $x \neq y$, we have $|x - y| \geqslant \dfrac{xy}{25}$. Find the maximum value of $|A|$. (The 26th IMO alternative)

Analysis. First we want to remove the absolute value symbol. This requires that we have an ordering of the elements, which leads to the idea of ordering the elements of the set A, and using space estimation to study the number of elements under the given conditions.

Solution. Let $A = \{a_1 < a_2 < \cdots < a_n\}$, then the condition implies that for any $i < j$, we have $a_j - a_i \geqslant \dfrac{a_j a_i}{25}$. Then $a_{i+1} - a_i \geqslant \dfrac{a_i a_{i+1}}{25}$, thus

$$\frac{1}{a_i} - \frac{1}{a_{i+1}} \geqslant \frac{1}{25}.$$

Letting $i = 1, 2, \ldots, n - 1$, and summing up the above inequalities, we have $\dfrac{1}{a_1} - \dfrac{1}{a_n} \geqslant \dfrac{n-1}{25}$. Therefore $\dfrac{1}{a_1} > \dfrac{n-1}{25}$, so $n - 1 < \dfrac{25}{a_1} \leqslant 25$, $n \leqslant 25$. However, this estimate is not sharp, since we can improve it by "*increasing the starting point*".

Letting $i = 2, 3, \ldots, n - 1$, we have $\dfrac{1}{a_2} - \dfrac{1}{a_n} \geqslant \dfrac{n-2}{25}$, thus $\dfrac{1}{a_2} > \dfrac{n-2}{25}$. So $2 \leqslant a_2 < \dfrac{25}{n-2}$, $n < \dfrac{29}{2}$, $n \leqslant 14$. By experiments, we see that this is still not sharp; thus we still need to improve the estimation.

Similarly, we have $3 \leqslant a_3 < \dfrac{25}{n-3}$, $n \leqslant 11$; $4 \leqslant a_4 < \dfrac{25}{n-4}$, $n \leqslant 10$; $5 \leqslant a_5 < \dfrac{25}{n-5}$, $n \leqslant 9$; $6 \leqslant a_6 < \dfrac{25}{n-6}$, $n \leqslant 10$. From this tendency, we guess that $n \leqslant 9$ is the best estimate.

Next we prove that there is a subset A with nine elements which satisfies the requirement. First, when $xy \leqslant 25$, $|x - y| \geqslant \dfrac{xy}{25}$ is trivially satisfied; so we can choose $1, 2, 3, 4, 5 \in A$. Then $6 \notin A$, since $\mid 6 - 5 \mid < \dfrac{30}{25}$, which does not satisfy the requirement. Going on like this we can obtain that 7, 10, 17, $54 \in A$. Thus, $A = \{1, 2, 3, 4, 5, 7, 10, 17, 54\}$ is the required subset.

Example 3. Let $X = \{1, 2, \ldots, n\}$, $A_i = \{a_i, b_i, c_i\}(i = 1, 2, \ldots, m\}$ be subsets of X, each of which has three elements. For any A_i, $A_j (1 \leqslant i < j \leqslant m)$, at most one of the three equations $a_i = a_j$, $b_i = b_j$, $c_i = c_j$ holds. Find the maximum value of m.

Solution. For $2 \leqslant k \leqslant n - 1$, consider the subsets $\{a_i, k, c_i\}$ with three elements, where k is the middle element, so that $a_i < k < c_i$. Let $f(k)$ be the number of such subsets. Since a_i can be chosen from 1, 2, $\ldots$, $k - 1$, we know $f(k) \leqslant k - 1$. Also since c_i can be chosen from n, $n - 1$, $\ldots$, $k + 1$, we have $f(k) \leqslant n - k$.

Thus $f(k) \leqslant \min\{k - 1, n - k\}$.

Then

$$m = \sum_{k=2}^{n-1} f(k) \leqslant \sum_{k=2}^{n-1} \min\{k - 1, n - k\} = \begin{cases} \dfrac{n(n-2)}{4} & (n \text{ is even}), \\[2mm] \left(\dfrac{n-1}{2}\right)^2 & (n \text{ is odd}). \end{cases}$$

Now let A_1, A_2, $\ldots$, A_m be all the subsets $\{a, b, c\}$ which satisfy $a + c = 2b$, then

$$m = \begin{cases} \dfrac{n(n-2)}{4} & (n \text{ is even}), \\[2mm] \left(\dfrac{n-1}{2}\right)^2 & (n \text{ is odd}). \end{cases}$$

Thus

$$m_{\max} = \begin{cases} \dfrac{n(n-2)}{4} & (n \text{ is even}), \\[2mm] \left(\dfrac{n-1}{2}\right)^2 & (n \text{ is odd}). \end{cases}$$

Example 4. Let $a_1 < a_2 < \cdots < a_n = 100$ be positive integers. For any $i \geqslant 2$, there exist $1 \leqslant p \leqslant q \leqslant r \leqslant i-1$, such that $a_i = a_p + a_q + a_r$. Find the maximum and minimum value of n. (Original)

Solution. (1) Obviously, $n \neq 1$, 2. Thus $n \geqslant 3$.

When $n = 3$, let $a_1 = 20$, $a_2 = 60$, $a_3 = 100$, then $a_2 = a_1 + a_1 + a_1$, $a_3 = a_1 + a_1 + a_2$, so $n = 3$ satisfies the requirement. Thus the minimum value of n is 3.

(2) If $a_1 \equiv 1 \pmod 2$, then $a_2 = 3a_1 \equiv 3 \equiv 1 \pmod 2$. If for $i \leqslant k (k \geqslant 2)$, we have $a_k \equiv 1 \pmod 2$, then since $a_{k+1} = a_p + a_q + a_r (p \leqslant q \leqslant r \leqslant k)$, and by assumption $a_p \equiv 1 \pmod 2$, $a_q \equiv 1 \pmod 2$, $a_r \equiv 1 \pmod 2$, we have $a_{k+1} = a_p + a_q + a_r \equiv 1 + 1 + 1 \equiv 1 \pmod 2$. Thus for all $i = 1, 2, \ldots, n$, we have $a_i \equiv 1 \pmod 2$, which contradicts the condition $100 \equiv 0 \pmod 2$. Thus $a_1 \equiv 0 \pmod 2$, thus $a_1 \equiv 0$, 2 $\pmod 4$.

If $a_1 \equiv 2 \pmod 4$, then $a_2 = 3a_1 \equiv 6 \equiv 2 \pmod 4$.

If for $i \leqslant k (k \geqslant 2)$, $a_k \equiv 2 \pmod 4$, then since $a_{k+1} = a_p + a_q + a_r (p \leqslant q \leqslant r \leqslant k)$, and by assumption $a_p \equiv 2 \pmod 4$, $a_q \equiv 2 \pmod 4$, $a_r \equiv 2 \pmod 4$, we have $a_{k+1} = a_p + a_q + a_r \equiv 2 + 2 + 2 \equiv 2 \pmod 4$. Thus for all $i = 1, 2, \ldots, n$, we have $a_i \equiv 2 \pmod 4$, which contradicts the fact that $100 \equiv 0 \pmod 4$. Thus $a_1 \equiv 0 \pmod 4$, thus $a_1 \equiv 0$, 4 $\pmod 8$.

If $a_1 \equiv 0 \pmod 8$, then $a_2 = 3a_1 \equiv 0 \pmod 8$.

If for $i \leqslant k (k \geqslant 2)$, $a_k \equiv 0 \pmod 8$, then since $a_{k+1} = a_p + a_q + a_r$, $(p \leqslant q \leqslant r \leqslant k)$, and by assumption $a_p \equiv 0 \pmod 8$, $a_q \equiv 0 \pmod 8$, $a_r \equiv 0 \pmod 8$, we have $a_{k+1} = a_p + a_q + a_r \equiv 0 + 0 + 0 \equiv 0 \pmod 8$. Thus for all $i = 1, 2, \ldots, n$, we have $a_i \equiv 0 \pmod 8$, which contradicts the fact that $100 \equiv 4 \pmod 8$. Thus $a_1 \equiv 4 \pmod 8$, thus $a_1 \geqslant 4$.

Since $a_1 \equiv 4 \pmod 8$, we have $a_2 = 3a_1 \equiv 12 \equiv 4 \pmod 8$.

If for $i \leqslant k (k \geqslant 2)$, $a_k \equiv 4 \pmod 8$, then $a_{k+1} = a_p + a_q + a_r$, $(p \leqslant q \leqslant r \leqslant k)$, and by assumption $a_p \equiv 4 \pmod 8$, $a_q \equiv 4 \pmod 8$, $a_r \equiv 4 \pmod 8$, we have $a_{k+1} = a_p + a_q + a_r \equiv 4 + 4 + 4 \equiv 4 \pmod 8$. Thus for all $i = 1, 2, \ldots, n$, we have $a_i \equiv 4 \pmod 8$, so when $i \geqslant 2$, we have $a_i - a_{i-1} \equiv 0 \pmod 8$. So $a_i - a_{i-1} \geqslant 8$ (space estimation), or

$a_i \geqslant a_{i-1} + 8.$

Therefore, we have $a_n \geqslant a_{n-1} + 8 \geqslant a_{n-2} + 2 \times 8 \geqslant a_{n-3} + 3 \times 8 \geqslant \cdots \geqslant a_1 + (n-1) \times 8 \geqslant 4 + (n-1) \times 8 = 8n - 4$, thus $8n \leqslant a_n + 4 = 104$, so $n \leqslant 13$.

When $n = 13$, let $a_i = 8i - 4 (i = 1, 2, \ldots, 13)$, then for $i \geqslant 2$, we have $a_i = a_{i-1} + 8 = a_{i-1} + a_1 + a_1$.

Thus $n = 13$ satisfies the requirement, thus the maximum value of n is 13.

In conclusion, n has a minimum value of 3 and a maximum value of 13.

Note that n has many possible values. For example when $n = 5$, 4, 12, 36, 44, 100 satisfy the requirement. Actually, $12 = 4 + 4 + 4$, $36 = 12 + 12 + 12$, $44 = 36 + 4 + 4$, $100 = 44 + 44 + 12$.

Exercise 6

(1) Assume that X is a subset of $\mathbf{N}^*$. The minimum element of X is 1, and the maximum element is 100. For any element in X which is larger than 1, it can be written as the sum of two elements in X (which can be the same). Find the minimum value of $|X|$.

(2) In an $n \times n$ chessboard C, two grids are called "*connected*" if they share a common vertex. Write 1, 2, 3, $\ldots$, n^2 in the grids, such that each grid contains one number. If the difference between any two connected grids is at most g, then g is called a C-gasp. Find the minimum C-gasp C_g. (The 42nd Putanam Match Competition)

(3) Assume that 2005 segments are joined to form a closed path, such that any two parts of the path do not lie on the same straight line. What is the maximum number of self-intersection points of this path?

(4) Assume that 5-element subsets A_1, A_2, $\ldots$, A_k of set $M = \{1, 2, \ldots, 10\}$ satisfy the following condition: any two elements in M appears in at most two subsets A_i, $A_j (i \neq j)$. Find the maximum value of k. (2003 China Mathematical Olympiad Practice)

Chapter 7   Block Estimates

To estimate the number of elements in a subset A of set X which has some property, we can divide X into some blocks X_1, X_2, ..., X_t, and then discuss the number of elements of A in each block X_i, thus estimate the range of $|A|$. Usually we have the following three cases:

Case 1. Suppose that the subset A of X satisfies that: any r-element group of A has property p, and we want to find the maximum value of $|A|$. Then we can divide X into some blocks X_1, X_2, ..., X_t, so that any r-element group in X_i does not have property p. Thus there are at most $r-1$ elements of A in each X_i.

Case 2. Let $X = X_1 \cup X_2 \cup \cdots \cup X_t$, and each X_i has at most k_i elements of A. Thus $|A| \leqslant k_1 + k_2 + \cdots + k_t$. Obviously, the smaller $k_1 + k_2 + \cdots + k_t$ is, the more precise the estimation is (the more likely equality will hold). Thus, to divide X, we should let $k_1 + k_2 + \cdots + k_t$ as small as possible. That is to say, the ratio of k_i in A_i, $\dfrac{k_i}{|A_i|}$ should be as small as possible. Usually the most precise estimation can be found by experiments.

Case 3. Some properties are translation invariant, namely: if set $A = \{a_1, a_2, \ldots, a_n\}$ has property p, then $A + a = \{a_1 + a, a_2 + a, \ldots, a_n + a\}$ also has property p. Then, we can divide X evenly (the number of elements in each block being equal to each other), and then estimate each block.

Example 1. Let $M = \{1, 2, \ldots, 2005\}$, and A be a subset of M. If for any a_i, $a_j \in A$ with $a_i \neq a_j$, exactly one isosceles triangle can be determined with a_i, a_j as its side lengths, find the maximum value of $|A|$.

This problem has been solved in the previous chapter by space estimation. Now we solve it using block estimation. The basic idea is to divide M into some blocks, such that A has at most one element in each block. Note that the condition A satisfies is that: for any $a_i < a_j \in A$, we have $2a_i \leqslant a_j$. Thus, to divide M, we should let any two elements x, $y(x < y)$ in each block satisfy $2x > y$.

Solution. Divide M into 11 subsets: $A_1 = \{1\}$, $A_2 = \{2, 3\}$, $A_3 = \{2^2, 2^2 + 1, \ldots, 2^3 - 1\}, \ldots, A_{11} = \{2^{10}, 2^{10} + 1, \ldots, 2005\}$. Since for any two elements x, y in $A_i (x < y)$, we have $2x > y$, we know that $|A \cap A_i| \leqslant 1 \ (i = 1, 2, 3, \ldots, 11)$. So $|A| \leqslant 11$. Since $A = \{1, 2, \ldots, 1024\}$ satisfies the requirement, the maximum value of $|A|$ is 11.

Example 2. Let A be a subset of the set $\mathbf{N}^*$ of positive integers. For any x, $y \in A$, $x \neq y$, we have $|x - y| \geqslant \dfrac{xy}{25}$. Find the maximum value of $|A|$. (The 26th IMO alternative)

Analysis. This problem is solved in the previous chapter using space estimation. Now we solve it using block estimation. The basic idea is to divide $\mathbf{N}^*$ into several blocks, such that A has at most one element in each block. Note that A is such that for any $a_i < a_j \in A$, we have $a_j - a_i \geqslant \dfrac{a_i a_j}{25}$. Thus, to divide $\mathbf{N}^*$, we should let any two elements x, $y(x < y)$ in each block satisfy $y - x < \dfrac{xy}{25}$.

Solution. Let $X_1 = \{1\}$, $X_2 = \{2\}$, $X_3 = \{3\}$, $X_4 = \{4\}$, $X_5 = \{5, 6\}$, $X_6 = \{7, 8, 9\}$, $X_7 = \{10, 11, \ldots, 16\}$, $X_8 = \{17, 18, \ldots, 53\}$, $X_9 = \{54, 55, \ldots\} = \mathbf{N}^* \setminus \{1, 2, \ldots, 53\}$.

For X_9, when x, $y \in X_9$, $x > 25$, then $y - x < y < y \cdot \dfrac{x}{25} = \dfrac{xy}{25}$.

For $X_i (i = 1, 2, \ldots, 8)$, when x, $y \in X_i$, obviously $y - x < \dfrac{xy}{25}$.

Thus A contains at most one element in each block, so $n \leqslant 9$.

Also $A = \{1, 2, 3, 4, 5, 7, 10, 17, 54\}$ satisfies the requirement, so $|A|$ has a maximum value of 9.

Example 3. Let $A \subseteq \{0, 1, 2, \ldots, 29\}$ be such that, for any integer k and any elements a, b of A (possibly $a = b$), $a + b + 30k$ is not the product of two adjacent integers. Find all sets A which have the maximum number of elements. (2003 Chinese National Team Selection)

Solution. The desired subset $A = \{3t + 2 \mid 0 \leqslant t \leqslant 9, t \in \mathbf{Z}\}$.

Assume that A satisfies the requirements and $\mid A \mid$ is maximized. For two adjacent integers a, $a + 1$, we have $a(a + 1) \equiv 0, 2, 6, 12, 20, 26 \pmod{30}$; for any $a \in A$, let $b = a$, $k = 0$, we have $2a \not\equiv 0, 2, 6, 12, 20, 26 \pmod{30}$, thus $a \not\equiv 0, 1, 3, 6, 10, 13, 15, 16, 18, 21, 25, 28 \pmod{30}$. Thus $A \subseteq M = \{2, 4, 5, 7, 8, 9, 11, 12, 14, 17, 19, 20, 22, 23, 24, 26, 27, 29\}$, and M can be divided into the following 10 subsets, $A_1 = \{2, 4\}$, $A_2 = \{5, 7\}$, $A_3 = \{8, 12\}$, $A_4 = \{11, 9\}$, $A_5 = \{14, 22\}$, $A_6 = \{17, 19\}$, $A_7 = \{20\}$, $A_8 = \{23, 27\}$, $A_9 = \{26, 24\}$, $A_{10} = \{29\}$, so that each subset A_i contains at most one element in A. This means that $\mid A \mid \leqslant 10$.

If $\mid A \mid = 10$, then each subset A_i contains exactly one element of A. Thus $20 \in A$, $29 \in A$. From $20 \in A$ we can see that $12 \notin A$, $22 \notin A$, thus $8 \in A$, $14 \in A$, thus $4 \notin A$, $24 \notin A$. So $2 \in A$, $26 \in A$. Since $29 \in A$, we have $7 \notin A$, $27 \notin A$, therefore $5 \in A$, $23 \in A$, and $9 \notin A$, $19 \notin A$. Thus $11 \in A$, $17 \in A$.

In conclusion, $A = \{2, 5, 8, 11, 14, 17, 20, 23, 26, 29\}$. We can check that A satisfies the requirements, so it is the unique solution.

Example 4. Let A be a subset of $X = \{1, 2, 3, \ldots, 1989\}$. For any x, $y \in A$, we have $\mid x - y \mid \neq 4, 7$. Find the maximum value of $\mid A \mid$. (7th America Mathematical Olympiad)

Analysis. Let A be a subset of $X = \{1, 2, 3, \ldots, 1989\}$. If for any x, $y \in A$, we have $\mid x - y \mid \neq 4, 7$, we call A a good subset. Apparently, the property of being "*good*" is translation invariant; that is, if A is a good subset, then for any a, $A + a$ is also a good set. Thus we can apply the block estimation with an even partition.

Let $X = P_1 \cup P_2 \cup \cdots \cup P_k$. Note that our goal is to prove $|A| \leqslant r$ for r as small as possible. So we should make the number of good elements (elements in A) as small as possible in P_i. Let $P_1 = \{1, 2, \ldots, t\}$. This information is studied in the following table:

$t = \|P\|$	1	2	3	4	5	6	7	8	9	10	11	12	13
$\|A \cap P\|$	1	2	3	4	4	4	4	4	5	5	5	6	7
$\dfrac{\|A \cap P\|}{\|P\|}$	1	1	1	1	$\dfrac{4}{5}$	$\dfrac{2}{3}$	$\dfrac{4}{7}$	$\dfrac{1}{2}$	$\dfrac{5}{9}$	$\dfrac{1}{2}$	$\dfrac{5}{11}$	$\dfrac{1}{2}$	$\dfrac{7}{13}$

When $P = \{1, 2, \ldots, 11\}$, we get the minimum ratio $\dfrac{5}{11}$, so we guess that the division with $\{1, 2, \ldots, 11\}$ as a subset is the best.

Solution. Let A be such a subset. For $P = \{1, 2, \ldots, 11\}$, we prove that $|A \cap P| \leqslant 5$.

Divide P into six subsets $\{1, 5\}$, $\{2, 9\}$, $\{3, 7\}$, $\{4, 8\}$, $\{6, 10\}$, $\{11\}$. For each subset, A contains at most one element in it. So $|A \cap P| \leqslant 6$. If $|A \cap P| = 6$, then A has exactly one element in each subset, thus $11 \in A$, $\Rightarrow 4 \notin A$, $\Rightarrow 8 \in A$, $\Rightarrow 1 \notin A$, $\Rightarrow 5 \in A$, $\Rightarrow 9 \notin A$, $\Rightarrow 2 \in A$, $\Rightarrow 6 \notin A$, $\Rightarrow 10 \in A$, $\Rightarrow 3 \notin A$, $\Rightarrow 7 \in A$, but $11 - 7 = 4$, which is a contradiction. So $|A \cap P| \leqslant 5$.

Letting $P_k = \{11k + 1, 11k + 2, \ldots, 11k + 11\}$ ($k = 0, 1, 2, \ldots, 179$), $P_{180} = \{1981, 1982, \ldots, 1989\}$. We can see that A contains at most five elements in P_k ($k = 0, 1, 2, \ldots, 180$), so

$$|A| \leqslant 5 \times 181 = 905.$$

Finally, let $A_k = \{11k + 1, 11k + 3, 11k + 4, 11k + 6, 11k + 9\}$ ($k = 0, 1, 2, \ldots, 180$), $A = A_0 \cup A_1 \cup \cdots \cup A_{180}$, then A satisfies the requirement, and $|A| = 905$. Therefore the maximum value of $|A|$ is 905.

Example 5. Let p be a given positive integer, and A be a subset of $X = \{1, 2, 3, 4, \ldots, 2^p\}$ that has property: for any $x \in A$, we have $2x \notin A$. Find the maximum value of $|A|$. (1991 France Mathematical Olympiad)

Analysis and Solution. Divide X to blocks and induct on p.

When $p = 1$, $X = \{1, 2\}$, we may choose $A = \{1\}$, then $f(1) = 1$.

When $p = 2$, $X = \{1, 2, 3, 4\}$. Divide X into three subsets $\{1, 2\}$, $\{3\}$, $\{4\}$, then A contains at most one number in each subset, so $|A| \leqslant 3$. We may choose $A = \{1, 3, 4\}$, so $f(2) = 3$.

When $p = 3$, $X = \{1, 2, \ldots, 8\}$. Divide X into five subsets: $\{1, 2\}$, $\{3, 6\}$, $\{4, 8\}$, $\{5\}$, $\{7\}$, then A contains at most one number in each subset, thus $|A| \leqslant 5$. We may choose $A = \{1, 5, 6, 7, 8\}$, so $f(3) = 5$.

In general, when $X_p = \{1, 2, 3, \ldots, 2^p\}$, we can apply block estimation. Notice that A can contain $2^{p-1} + 1$, $2^{p-1} + 2$, $\ldots$, 2^p; this leads us to consider the following blocks, where $X = \{1, 2, 3, \ldots, 2^{p-1}\} \cup \{2^{p-1} + 1, 2^{p-1} + 2, \ldots, 2^p\} = X_{p-1} \cup M$, where $X_{p-1} = \{1, 2, 3, \ldots, 2^{p-1}\}$, $M = \{2^{p-1} + 1, 2^{p-1} + 2, \ldots, 2^p\}$. Then the problem turns into determining how many elements in $X_{p-1} = \{1, 2, 3, \ldots, 2^{p-1}\}$ can belong to A. Is this just the original problem in case $p - 1$? Not quite! Think about this: when the numbers $2^{p-1} + 1$, $2^{p-1} + 2$, $\ldots$, 2^p in M belong to A, there are many numbers in X_{p-1} cannot belong to A, such as $2^{p-2} + 1$, $2^{p-2} + 2$, $\ldots$, 2^{p-1}. However, we may not have that all of $\{2^{p-1} + 2, 2^{p-1} + 4, \ldots, 2^p\}$ belong to a. Therefore, the division should be made more precise: the numbers $2^{p-1} + 2$, $2^{p-1} + 4$, $\ldots$, 2^p in M should be related to certain numbers in X_{p-1} in terms of pairings: $\{2^{p-1} + 2, 2^{p-2} + 1\}$, $\{2^{p-1} + 4, 2^{p-2} + 2\}$, $\ldots$, $\{2^p, 2^{p-1}\}$. Then we have a division of X_p: $X_{p-2} = \{1, 2, 3, \ldots, 2^{p-2}\}$, $M_t = \{2^{p-1} + 2t, 2^{p-2} + t\}(t = 1, 2, \ldots, 2^{p-2})$, $M_0 = \{2^{p-1} + 1, 2^{p-1} + 3, 2^{p-1} + 5, \ldots, 2^{p-1} + 2^{p-1} - 1\}$.

Since A contains at most one element in each $M_t (t = 1, 2, \ldots, 2^{p-2})$, and contains at most $f(p - 2)$ elements in X_{p-2}, and contains at most 2^{p-2} elements in M_0, we have

$$f(p) \leqslant f(p - 2) + 2^{p-2} + 2^{p-2} = f(p - 2) + 2^{p-1}.$$

Next consider whether we can construct a set A, such that

$$f(p) \geqslant f(p - 2) + 2^{p-1}.$$

Assume that the subset of $X = \{1, 2, 3, \ldots, 2^{p-2}\}$ that satisfies the assumption and has the maximal number of elements is A_1, let $A_2 = \{2^{p-1} + 1, 2^{p-1} + 2, \ldots, 2^p\}$, then for any $x \in A_1$, we have $2x \leq 2 \cdot 2^{p-2} = 2^{p-1} < 2^{p-1} + 1 \notin A_2$. Thus $A = A_1 \cup A_2$ satisfies the requirements, so $f(p) \geq |A| = f(p-2) + 2^{p-1}$.

In conclusion, $f(p) = f(p-2) + 2^{p-1}$.

Next we solve the recursion relation using two different methods.

Method 1. Iterating (summing up the p equations), we have

$$f(p-1) + f(p) = f(1) + f(2) + 2^2 + 2^3 + \cdots + 2^{p-1}$$
$$= 1 + (2^0 + 2^1) + 2^2 + 2^3 + \cdots + 2^{p-1} = 2^p.$$

Iterating again (subtracting the $(p-2)$nd equation from the $(p-1)$th equation, then adding the $(p-3)$rd equation, and so on), we have

$$f(p) + (-1)^p f(1) = 2^p - 2^{p-1} + \cdots + (-1)^p \cdot 2^2,$$

so $$f(p) = 2^p - 2^{p-1} + \cdots + (-1)^p \cdot 2^2 + (-1)^{p+1}.$$

Note that

$$(-1)^{p+1} \cdot 2^1 + (-1)^{p+2} \cdot 2^0 = (-1)^{p+1}(2-1) = (-1)^{p+1},$$

so

$$f(p) = 2^p - 2^{p-1} + \cdots + (-1)^p \cdot 2^2 + (-1)^{p+1} \cdot 2^1 + (-1)^{p+2} \cdot 2^0$$
$$= \frac{2^p \left[1 - \left(-\frac{1}{2}\right)^{p+1}\right]}{1 + \frac{1}{2}} = \frac{2^{p+1} + (-1)^p}{3}.$$

Method 2. Solve the problem in different cases separately.

When p is odd,

$$f(p) = f(p-2) + 2^{p-1} = f(p-4) + 2^{p-3} + 2^{p-1}$$
$$= f(1) + 2^2 + 2^4 + \cdots + 2^{p-1}$$
$$= 2^0 + 2^2 + 2^4 + \cdots + 2^{p-1} = \frac{2^{p+1} - 1}{3}.$$

When p is even,

$$f(p) = f(p-2) + 2^{p-1} = f(p-4) + 2^{p-3} + 2^{p-1}$$

$$= f(2) + 2^3 + 2^5 + \cdots + 2^{p-1}$$
$$= 1 + 2^1 + 2^3 + 2^5 + \cdots + 2^{p-1} = \frac{2^{p+1} + 1}{3}.$$

Example 6. Let $X = \{1, 2, \ldots, 2001\}$. Find the minimal positive integer m, such that for any subset W of X with m elements, there exist $u, v \in W$ (allowing $u = v$), such that $u + v$ is powers of 2. (2001 China Mathematical Olympiad)

Analysis and Solution. For simplicity, if $u + v$ is powers of 2, then (u, v) is called a pair. Assume the contrary. If W does not contain any pair, how many elements can W have? Apparently, if X can be divided into blocks, where any two numbers in each block form a pair, then W can only contain one element in each block. Thus, let $A_i = \{1024 - i, 1024 + i\}$ ($i = 1, 2, \ldots, 977$), $B_j = \{32 - j, 32 + j\}$ ($j = 1, 2, \ldots, 14$), $C = \{15, 17\}$, $D_k = \{8 - k, 8 + k\}$ ($k = 1, 2, \ldots, 6$), $E = \{1, 8, 16, 32, 1024\}$. If W does not contain any pair, then W cannot contain any element in E, and can contain at most one element in each A_i, B_j, D_k and C. Therefore $|W| \leqslant 977 + 14 + 6 + 1 = 998$. This means that when $|W| \geqslant 999$, W contains at least one pair, so $m = 999$ satisfies the requirement. Next, if $|W| = 998$ and W does not contain any pair, then W contains exactly one element in each A_i, B_j, D_k and C. Let $W = \{1025, 1026, \ldots, 2001\} \cup \{33, 34, \ldots, 46\} \cup \{17\} \cup \{9, 10, 14\}$, it is easy to verify that W contains no pair. Thus, when $m < 999$, we can choose a subset of W with m elements, so that it has no pair. Therefore $m \geqslant 999$.

In conclusion, the minimum value of m is 999.

Example 7. Let n be a fixed even positive integer. Consider an $n \times n$ square chessboard. Two grids are called "neighboring" if they share a common edge.

Now, mark N grids in the chessboard, such that any grid in the chessboard (marked or unmarked) is the neighbor of a marked grid.

Find the minimum value of N. (40th IMO)

Solution. If n is not divided by 4, we paint the chessboard as in Figure 7.1; otherwise paint the chessboard as in Figure 7.2. Consider all black grids. If $n = 4k$, after painting as Figure 7.2, there will be $4 \times 3 + 4 \times 7 + \cdots + 4 \times (4k - 1) = 2k(4k + 2)$ black grids. If $n = 4k + 2$, after painting as in Figure 7.1, there will be

$$4 \times 1 + 4 \times 5 + \cdots + 4 \times (4k + 1) = 2(k + 1)(4k + 2)$$

black grids. Either way, there are $\frac{1}{2}n(n + 2)$ black grids, and any three of them do not have a common neighbor. From the requirements, each of them should have a marked neighbor. So $N \geqslant \frac{1}{4}n(n + 2)$.

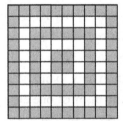   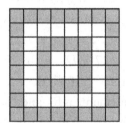

Figure 7.1   **Figure 7.2**

1	2	3							n
$4n{-}4$									$n{+}1$
									$n{+}2$
$3n{-}2$									$2n{-}1$

Figure 7.3

Next we prove that, we can mark $N = \frac{1}{4}n(n+2)$ grids so that the requirement is met.

Actually, as in Figure 7.3, we number the $4(n-1)$ grids of the "first layer boundary" of the chessboard with $1, 2, \ldots, 4n-4$ counterclockwise, starting from the top left corner. We then mark the grids with numbers congruent to 1 or 2 mod 4 (the shadow grids in Figure 7.3).

We remove the two outer layers of the boundary, and number the remaining chessboard in the similar (clockwise) way, and again mark the grids with numbers congruent to 1 or 2 mod 4. This goes on until the process is finished. In this way we have marked half of the black grids, so there are $\frac{1}{4}n(n+2)$ grids marked. Next we prove that this satisfies the requirement.

Actually, from Figure 7.3, it is clear that any two marked grids do not share any neighbor. Assume that the marked grids are $A_1, \ldots, A_N$, where $N = \frac{1}{4}n(n+2)$, and the set of neighbors of A_i is M_i. Then $M_i (i = 1, 2, \ldots, N)$ do not intersect with each other. For grids A_i lying in a corner (there are two such grids), $|M_i| = 2$. For grids A_i lying on the boundary of the chessboard (there are $2n-4$ such grids), $|M_i| = 3$. For grids A_i not on the boundary (there are $\frac{n^2-6n+8}{4}$ such grids), $|M_i| = 4$. Therefore

$$|M_1 \cup M_2 \cup \cdots \cup M_N| = \frac{n^2-6n+8}{4} \times 4 + (2n-4) \times 3 + 2 \times 2 = n^2.$$

So $M_1 \cup M_2 \cup \cdots \cup M_N$ contains all grids. Thus, each grid has a marked neighbor.

In conclusion, $N_{\min} = \frac{1}{4}n(n+2)$.

Example 8. There are 2000 points in the xOy plane, forming a point set S. We know that the line connecting any two points is not parallel to the axes. For any two points P, Q in S, consider the rectangle M_{PQ} whose diagonal is PQ and whose edges are parallel to the axes. We use W_{PQ} to represent the number of points in rectangle M_{PQ}.

If the statement "no matter how the points in S distribute in the plane, there are at least a pair of points P, Q, which makes $W_{PQ} \geqslant N$" is true, find the maximum value of N. (2002 Japan Mathematical Olympiad 2nd Round)

Solution. $N_{\max} = 400$. First we prove that there exist points P, Q which make $W_{PQ} \geqslant 400$. Actually, let A be the point in S whose y-coordinate is maximal, B be the point whose y-coordinate is minimal, C be the point whose x-coordinate is maximal, D be the point whose x-coordinate is minimal. If two points in A, B, C, D coincide, then the conclusion obviously holds. (Assume $A = C$, then in this case M_{AB}, M_{BD}, M_{AD} cover S. So $\max\{W_{AB}, W_{BD}, W_{AD}\} \geqslant \dfrac{2002 - 3}{3} > 400$). If no two of the points in A, B, C, D coincide, then the distribution will be like Figure 7.4.

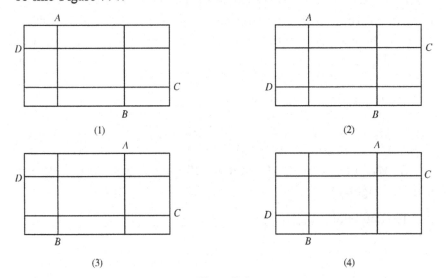

(1)

(2)

(3)

(4)

Figure 7.4

In case (1), (4), M_{AC}, M_{BC}, M_{BD} cover S, so $\max\{W_{AC}, W_{BD},$

$W_{AD}, W_{BC}\} \geqslant \dfrac{2002 - 4}{4} > 400$.

In case (2), (3), M_{AC}, M_{BC}, M_{AD}, M_{BD}, M_{AB} cover S, so

$\max\{W_{AC}, W_{BD}, W_{AD}, W_{BC}, W_{AB}\} \geqslant \dfrac{2002 - 4}{5} > 399$. Thus $\max\{W_{AC},$

$W_{BD}, W_{AD}, W_{BC}, W_{AB}\} \geqslant 400$. Therefore there exist P, Q, which make $W_{PQ} \geqslant 400$.

Next we prove that there exists an S, such that for all P, $Q \in S$, we have $W_{PQ} \leqslant$ 400. As Figure 7.5 shows, the 2002 points in S are divided into five groups E, F, G, H, I, where H and F have 401 points, and E, I, G have 400 points, and the points in each group lie on the diagonal of the corresponding rectangle. Apparently, any rectangle M_{PQ} (P, $Q \in S$) contains points

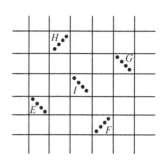

Figure 7.5

from at most one group. If it contains group I, then $W_{PQ} \leqslant 400$. If it contains any other group, then at least one point of P, Q belongs to this group. Thus

$$W_{PQ} \leqslant 401 - 1 = 400.$$

In conclusion, $N_{\max} = 400$.

Example 9. In a 7×8 chessboard, each grid is placed with a stone. If two grids share a vertex, then the two stones placed on these two grids are called connected. Now remove r stones from the chessboard, so that there are no five stones connected one by one in a same straight line (horizontal, vertical or in diagonal direction). Find the minimum value of r. (2007 China National Mathematical Competition)

Solution. We call a grid with stone removed a "blank". Assume that there are k blanks in the chessboard.

As Figure 7.6 shows, four straight lines divide the chessboard into nine blocks A, B, C, D, E, F, G, H, O.

From the condition, $A \cup G$ has at least two blanks, $E \cup O$ has at least three blanks, $C \cup H$ has at least two blanks, $D \cup F$ has at least three blanks, thus $k \geqslant 2 + 3 + 2 + 3 = 10$.

If $k = 10$, then area D has no blank. By symmetry, each of A, B, C, D has no blank. Thus, since $A \cup G$ has at least two blanks and A has no blank, G must have at least two blanks. Similarly, H has at least two blanks. Since $A \cup E$ has at least three blanks and A has no blank, we know that E has at least three blanks. Similarly, F has at least three blanks. So E, F, G, H have at least ten blanks in total. Since the chessboard has only ten blanks, we know that E, F have only three blanks, G, H have just two blanks, O has no blank.

Since $A \cup E$ has at least one blank in each column and A has no blank, we know that E has at least one blank in each column. Since E has only three blanks, we know that E has only one blank in each column. Then at least one of grids 1, 2 is not blank. Thus, at least one straight line in Figure 7.6 contains five connected stones, which is a contradiction. So $k \geqslant 11$.

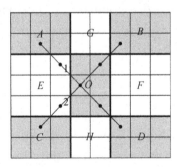

Figure 7. 6

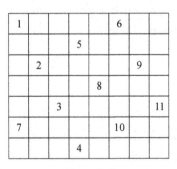

Figure 7. 7

When $k = 11$, as Figure 7.7 shows, we can place the stones using horse steps, so that the chessboard does not have five stones connected in a straight line. Note that this seems more natural than the official solution.

In conclusion, $k_{\min} = 11$.

Example 10. Let the set $S = \{1, 2, \ldots, 50\}$, and X be any subset of S where $|X| = n$. Find the minimum value of n, so that X always has three numbers which form the side lengths of a right-angled triangle.

Solution. Let the side lengths of a right-angled triangle be x, y, z. We have $x^2 + y^2 = z^2$, the solution in positive integers being

$$x = k(a^2 - b^2), \; y = 2kab, \; z = k(a^2 + b^2), \qquad \text{①}$$

where k, a, $b \in \mathbf{N}^*$ and $(a, b) = 1$, $a > b$.

First, one of x, y, z should be multiple of 5. In fact, if a, b, c are not multiples of 5, then a, b, c are numbers of the form $5m \pm 1$, $5m \pm 2 (m \in \mathbf{N})$, thus $a^2 \equiv \pm 1 \pmod 5$, $b^2 \equiv \pm 1 \pmod 5$, $c^2 \equiv \pm 1 \pmod 5$, hence $c^2 = a^2 + b^2 \equiv 0$ or ± 2, which is a contradiction.

Let the set $A = \{$numbers in S which are relatively prime to $5\}$; then Card $A = 40$. If we use 10, 15, 25, 40, 45 as one side length of the right-angled triangle, then from ①, we can find numbers in A to form the following right-angled triangles: $(10, 8, 6)$, $(26, 24, 10)$, $(15, 12, 9)$, $(17, 15, 8)$, $(39, 36, 15)$, $(25, 24, 7)$, $(40, 32, 24)$, $(41, 40, 9)$, $(42, 27, 36)$, and these are the only possibilities if one side length of the right-angles triangle is 10, 15, 25, 40, 45 and the other two are in A.

Let $M = A \cup \{10, 15, 25, 40, 45\} \setminus \{8, 9, 24, 36\}$; then Card $M = 41$.

We can see that any three numbers in A cannot form a right-angled triangle. Since M does not contain 8, 9, 24, 36, we cannot find any numbers in M that form a right-angled triangle with 10, 15, 25, 40, 45. Thus any three numbers in M cannot form a right-angled triangle, thus $n \geqslant 42$.

On the other hand, from ①, we can choose the integer sets: $B = \{3, 4, 5, 17, 15, 8, 29, 21, 20, 25, 24, 7, 34, 16, 30, 37, 35, 12, 50, 48, 14, 41, 40, 9, 45, 36, 27\}$, where any three numbers with the same underline can form a right-angled triangle, and we have Card $B = 27$.

The number of elements in $S \setminus B$ is $50 - 27 = 23$. For any 42 numbers in S, since $42 - 23 = 19$, there must be 19 numbers among the

42 chosen numbers that come from B. So there must be three numbers among the 42 numbers that come from the same underlined group. Therefore, for any 42 numbers, there are at least three numbers which can form a right-angled triangle. So the minimum value of n is 42.

Exercise 7

(1) Let $X = \{1, 2, 3, \ldots, 1993\}$, and A be a subset of X, which satisfies: (i) for any two numbers $x \neq y$ in A, 93 cannot divide $x \pm y$; (ii) $S(A) = 1993$. Find the maximum value of $|A|$.

(2) Let $X = \{1, 2, 3, \ldots, 10\}$, and A be a subset of X such that for any $x < y < z$, x, y, $z \in A$, there exists a triangle whose side lengths are x, y, z. Find the maximum value of $|A|$.

(3) Let $X = \{1, 2, 3, \ldots, 20\}$, and A be a subset of X such that for any $x < y < z$, x, y, $z \in A$, there exists a triangle whose side lengths are x, y, z. Find the maximum value of $|A|$.

(4) Let $X = \{00, 01, \ldots, 98, 99\}$ be the set of 100 2-digits codes, and A be a subset of X such that for any infinite sequence composed of numbers between 0 to 9, there are two consecutive digits forming a 2-digit code that belongs to A. Find the minimum value of $|A|$. (52nd Moscow Mathematical Olympiad)

(5) How many number can be chosen from 1, 2, $\ldots$, 20, such that any of them is not two times any other chosen number. How many ways are there to choose these numbers?

(6) A natural number k satisfies that: in 1, 2, $\ldots$, 1988, k different numbers can be chosen, such that the sum of any two of them cannot be divided by the difference of them. Find the maximum value of k. (26th Moscow Mathematical Olympiad)

(7) In a subset S of set $X = \{1, 2, \ldots, 50\}$, the sum of squares of any two elements cannot be divided by 7. Find the maximum value of $|S|$.

(8) Let $X = \{1, 2, \ldots, 1995\}$, and A be a subset of X, such that if $x \in A$, then $19x \notin A$. Find the maximum value of $|A|$.

Chapter 8 Guesses and Contradiction

Sometimes sets possessing property p are easy to construct, where we only need to collect all the elements having a certain property. In this case we can first construct a set possessing property p, and then guess that this set is the maximum one. An effective approach to prove this statement is by contradiction.

Assume that A is a set possessing some property p, we expect to prove that $|A| \leqslant r$. Assume that $|A| > r$, we want to show that there must be some special element in A, which makes A fail to possess property p. Here the pigeonhole principle is an important tool.

Example 1. If a set does not contain three numbers x, y, z such that $x + y = z$, then the set is called simple. Assume $M = \{1, 2, \ldots, 2n + 1\}$, A is a simple subset of M. Find the maximum value of $|A|$. (1982 Germany Mathematical Olympiad)

Analysis. Considering odd$+$odd$\neq$odd, we find that $A = \{1, 3, 5, \ldots, 2n + 1\}$ is a simple set, where $|A| = n + 1$. We guess that the maximum value of $|A|$ is $n + 1$. Here we need to prove that: if $|A| \geqslant n + 2$, then there must be three numbers x, y, z in A, such that $x + y = z$. Note that $x + y = z$ means that $x = z - y$, thus we can use the difference between elements to form "*new*" elements and apply the pigeonhole principle.

Solution. Let $A = \{1, 3, 5, \ldots, 2n + 1\}$, then A is simple and $|A| = n + 1$. Next we prove that for any simple subset A, $|A| \leqslant n + 1$. We proceed by contradiction. Assume $|A| > n + 1$. Then there are at least $n + 2$ elements in A, assume they are $a_1 < a_2 < \cdots < a_{n+2}$.

Method 1. Assume that A has p odd numbers $a_1 < a_2 < \cdots < a_p$ and $n + 2 - p$ even numbers $b_1 < b_2 < \cdots < b_{n+2-p}$. Note that there are $n + 1$ odd numbers and n even numbers in M, so $p \geqslant 2$.

Consider the numbers $a_2 - a_1 < a_3 - a_1 < \cdots < a_p - a_1$. They are all positive even numbers. Adding $b_1 < b_2 < \cdots < b_{n+2-p}$, there will be $(p - 1) + (n + 2 - p) = n + 1 > n$ positive even numbers. Using pigeonhole principle, there must be two elements that are equal, and they can only be some b_i and some $a_j - a_1$. Thus $a_1 + b_i = a_j$. A is not simple, which is a contradiction.

Method 2. Consider the $2n + 2$ elements $a_2 < a_3 < \cdots < a_{n+2}$, and $a_2 - a_1 < a_3 - a_1 < \cdots < a_{n+2} - a_1$, they are all positive integers which are no more than $2n + 1$. So there are at least two elements that are equal. Assume $a_i - a_1 = a_j$, then A is not simple, which is a contradiction.

Method 3. Consider the $2n + 3$ elements $a_1 < a_2 < a_3 < \cdots < a_{n+2}$ and $a_2 - a_1 < a_3 - a_1 < \cdots < a_{n+2} - a_1$, they are all positive integers no more than $2n + 1$. Note that $2n + 3 - (2n + 1) = 2$, there must be two pairs of elements that are equal: $a_i - a_1 = a_j$, $a_s - a_1 = a_t$. Now at least one of a_j, a_t is not a_1, thus A is not simple, which is a contradiction.

In conclusion, the maximum value of $|A|$ is $n + 1$.

Example 2. Let $X = \{1, 2, \ldots, 100\}$, and A be a subset of X. If for any two elements x, y of $A (x < y)$, we have $y \neq 3x$, find the maximum value of $|A|$.

Analysis. First we construct a set $|A|$ satisfying the requirements. A sufficient condition to guarantee $y \neq 3x$ is that y is not a multiple of 3. Thus any number which is not a multiple of 3 can belong to A. Furthermore, there can also be some multiples of 3 in A. Such multiples should satisfy two conditions: first, they are not three times any of the previous numbers, which is satisfied when they are multiples of 9; second, they are not three times of any number among themselves, so that we can choose 9, 18, 36, 45, 63, 72, 81, 90, 99 to be in A. Thus $|A| = 76$, and we can guess that the maximum value of $|A|$ is 76.

Solution. Let $A = \{3k+1 \mid k = 0, 1, 2, \ldots, 33\} \cup \{3k+2 \mid k = 0, 1, 2, \ldots, 32\} \cup \{9, 18, 36, 45, 63, 72, 81, 90, 99\}$. Then A satisfies the requirements and $|A| = 76$.

On the other hand, consider the 24 sets $A_k = \{k, 3k\} (k = 1, 2, 12, 13, \ldots, 33)$, which contain 48 different numbers. There are 52 numbers left in X besides these numbers. Consider each number among these 52 numbers as a set with a single element. With the previous 24 sets, there are $52+24 = 76$ sets in total. If $|A| > 76$, then at least two numbers in A fall into the same set, thus the larger one is three times the smaller one, which is a contradiction.

In conclusion, the maximum value of $|A|$ is 76.

Example 3. For a set M of numbers, define the sum of M as the sum of every element in M, denoted by $S(M)$. Assume that M is a set of positive integers no more than 15, and any two disjoint subsets of M have different sums, find the maximum value of $S(M)$.

Analysis. From the requirements, it is easy to construct the set M with the maximal $S(M)$: first we choose 15, 14, 13, so we cannot choose 12; then we choose 11, so we cannot choose 10 or 9. Finally we choose 8, and do not choose any other number. Then we have set $M = \{15, 14, 13, 11, 8\}$ with $|M| = 5$ and $S(M) = 61$. Then we guess that $S(M) \leqslant 61$. To prove the guess, it is intuitive to assume that $|M|$ is not too large. Then we can guess furthermore that $|M| \leqslant 5$. Here, we can use contradiction and the pigeonhole principle.

Solution. Let $M = \{15, 14, 13, 11, 8\}$, then $S(M) = 61$. Now we prove that: for any set M satisfying the requirements, we have $S(M) \leqslant 61$. To do this, we first prove that: $|M| \leqslant 5$. ①

Assume the contrary. Consider any subset A which has no more than four elements, then $S(A) \leqslant 15+14+13+12 = 53$, and we have $C_6^1 + C_6^2 + C_6^3 + C_6^4 = 56$ such subsets. So there must be two subsets A, B, such that $S(A) = S(B)$. Let $A' = A \backslash (A \cap B)$, $B' = B \backslash (A \cap B)$, then A', B' are disjoint and $S(A') = S(B')$, which is a contradiction.

So $|M| \leqslant 5$.

Now consider any M.

    (1) If $15 \notin M$, then from ①, we have $S(M) \leqslant 14 + 13 + 12 + 11 + 10 = 60$.

    (2) If $14 \notin M$, then from ①, we have $S(M) \leqslant 15 + 13 + 12 + 11 + 10 = 61$.

    (3) If $13 \notin M$, then since $15 + 11 = 14 + 12$, we have $M \neq \{15, 14, 12, 11, 10\}$, so from ①, we have $S(M) < 15 + 14 + 12 + 11 + 10 = 62$.

    (4) If $15, 14, 13 \in M$, then $12 \notin M$.

    (i) If $11 \in M$, then $10, 9 \notin M$. From ①, we have

$$S(M) \leqslant 15 + 14 + 13 + 11 + 8 = 61.$$

    (ii) If $11 \notin M$, from ①, we have

$$S(M) \leqslant 15 + 14 + 13 + 10 + 9 = 61.$$

In conclusion, $S(M) \leqslant 61$. So the maximum value of $S(M)$ is 61.

Example 4. In a math contest, there are n questions $(n \geqslant 4)$. Each question was solved by exactly four people. For any two questions, there is exactly one person who solved both of them.

If there are at least $4n$ people participating the contest, find the minimum value of n, such that there is always a person who has solved all of the questions. (The 15th Korean Mathematical Olympiad)

Analysis and Solution. First, when $4 \leqslant n \leqslant 13$, we can construct a counterexample. Actually, since the number of participants is no less than $4n \geqslant 16$, we may consider the following 13 sets:

$M_1 = \{1, 2, 3, 4\}$, $M_2 = \{1, 5, 6, 7\}$, $M_3 = \{1, 8, 9, 10\}$,
$M_4 = \{1, 11, 12, 13\}$, $M_5 = \{2, 5, 8, 11\}$, $M_6 = \{2, 6, 9, 12\}$,
$M_7 = \{2, 7, 10, 13\}$, $M_8 = \{3, 5, 9, 13\}$, $M_9 = \{3, 6, 10, 11\}$,
$M_{10} = \{3, 7, 8, 12\}$, $M_{11} = \{4, 5, 10, 12\}$, $M_{12} = \{4, 7, 9, 11\}$,
$M_{13} = \{4, 6, 8, 13\}$.

    For $4 \leqslant n \leqslant 13$, choosing n sets $M_i (1 \leqslant i \leqslant n)$, it is easy to verify that $|M_i| = 4$, and $|M_i \cap M_j| = 1$, but M_1, M_2, ..., M_n has no common element, which is a contradiction. So we guess that $n_{\min} = 14$.

Next we prove that, $n = 14$ satisfies the requirements. Actually, assume the set of people who solved the ith question is M_i, then for any $1 \leqslant i \leqslant 14$, $|M_i| = 4$, and $|M_i \cap M_j| = 1$. Let $M_1 = \{a, b, c, d\}$, then each of the other 13 sets contains at least one element in a, b, c, d, thus there must be one element (e. g. a), which appears at least four times in these 13 sets. Assume M_2, M_3, M_4, M_5 contain a. We then prove that each M_i contains a.

Assume the contrary. Suppose there is t ($6 \leqslant t \leqslant 14$), such that a is not contained in M_t, and M_t shares one element with each of M_1, M_2, M_3, M_4, M_5. Assume the common element between M_t and M_k ($1 \leqslant k \leqslant 5$) is a_k ($1 \leqslant k \leqslant 5$), then $a_k \neq a$. If there is i, j ($1 \leqslant i < j \leqslant 5$), such that $a_i = a_j$, then a, a_i belong to $M_i \cap M_j$, which contradicts $|M_i \cap M_j| = 1$. So a_1, a_2, a_3, a_4, a_5 are different from each other, but they all belong to M_t, thus $|M_t| \geqslant 5$, which is a contradiction. So a is contained in any M_t, thus the participant a solved all the problems, so the conclusion holds.

Example 5. Eighteen teams participate in a single round-robin tournament, where in each round 18 teams are divided into nine pairs, each pair playing once, and in the next round they are regrouped. There are 17 rounds in total, such that each teams play against each of the other 17 teams exactly once. If, after n rounds have been played, there are always four teams such that only one game has been played among them, find the maximum value of n. (2002 China Mathematical Olympiad)

Solution. Consider the following schedule:

   1. (1, 2)(3, 4)(5, 6)(7, 8)(9, 18)(10, 11)(12, 13)(14, 15) (16, 17);

   2. (1, 3)(2, 4)(5, 7)(6, 9)(8, 17)(10, 12)(11, 13)(14, 16) (15, 18);

   3. (1, 4)(2, 5)(3, 6)(8, 9)(7, 16)(10, 13)(11, 14)(12, 15)

(17, 18);

    4. (1, 5)(2, 7)(3, 8)(4, 9)(6, 15)(10, 14)(11, 16)(12, 17)(13, 18);

    5. (1, 6)(2, 8)(3, 9)(4, 7)(5, 14)(10, 15)(11, 17)(12, 18)(13, 16);

    6. (1, 7)(2, 9)(3, 5)(6, 8)(4, 13)(10, 16)(11, 18)(12, 14)(15, 17);

    7. (1, 8)(2, 6)(4, 5)(7, 9)(3, 12)(10, 17)(11, 15)(13, 14)(16, 18);

    8. (1, 9)(3, 7)(4, 6)(5, 8)(2, 11)(10, 18)(12, 16)(13, 15)(14, 17);

    9. (1, 10)(2, 3)(4, 8)(5, 9)(6, 7)(11, 12)(13, 17)(14, 18)(15, 16);

    10. (1, 11)(2, 12)(3, 13)(4, 14)(5, 15)(6, 16)(7, 17)(8, 18)(9, 10);

    11. (1, 12)(2, 13)(3, 14)(4, 15)(5, 16)(6, 17)(7, 18)(8, 10)(9, 11);

    12. (1, 13)(2, 14)(3, 15)(4, 16)(5, 17)(6, 18)(7, 10)(8, 11)(9, 12);

    $\vdots$

    17. (1, 18)(2, 10)(3, 11)(4, 12)(5, 13)(6, 14)(7, 15)(8, 16)(9, 17).

Call the first nine teams group A, and the last nine teams group B. After nine rounds, any two teams in the same group have played a game. So, for any four teams, they have played at least two games among them, so the requirement is not satisfied. If we reverse the above schedule, then after eight rounds, any two teams in the same group have played no game; each team has played eight games with teams from the other group, thus any four teams from the same group have played no game among them, and four teams from different groups have played at least two games among each them, so the requirement is not satisfied. Thus $n \leqslant 7$. When $n - 7$, assume the contrary that any four teams do not satisfy the requirements. Choose two teams A_1, A_2 which have played a game, then each of them have played six games against the other teams, so the two teams have played with at most 12 teams. Thus there are at least four teams B_1, B_2, B_3, B_4, such that they have not played with A_1, A_2. Consider the four teams A_1, A_2, B_i, B_j ($1 \leqslant i < j \leqslant 4$), by assumption, they have played at least two games among them, thus for any $1 \leqslant i < j \leqslant 4$, B_i have played with B_j. Since B_1, B_2 have played three games within $\{B_1, B_2, B_3, B_4\}$, we know that B_1, B_2 have played with four teams among the other 14 teams. Thus there are at least six teams C_1, C_2, C_3, C_4, C_5, C_6, such that they have not played with B_1, B_2. Similarly, for any $1 \leqslant i < j \leqslant 6$, C_i and C_j have played. Since each of C_1, C_2 has played five games within $\{C_1, C_2, \ldots, C_6\}$, we know that C_1, C_2 have played with two teams among the other 12 teams, thus there are at least eight teams D_1, D_2, $\ldots$, D_8, such that they have not played with C_1, C_2. Similarly, for any $1 \leqslant i < j \leqslant 8$, D_i and D_j have played. Thus, D_1, D_2 have not played with the other 10 teams. Since there are only seven rounds, there are at least two teams E_1, E_2 in the other 10 teams which have not played, thus D_1, D_2, E_1, E_2 have played only one game among them, which is a contradiction. In conclusion, the maximum value of n is 7.

Example 6. Find the maximum value of the number of elements of a set S which satisfies the following requirements:

(1) Each element in S is a positive integer no more than 100.

(2) For any two different elements a, b in S, there is an element c in S, such that the greatest common divisor of a and c is 1, and the greatest common divisor of b and c is 1.

(3) For any two different elements a, b in S, there is a different element d, such that the greatest common divisor of a and d is larger than 1, and the greatest common divisor of b and d is larger than 1.
(2003 China Mathematical Olympiad)

Analysis and Solution. Represent a positive integer n no more than 100 as $n = 2^{\alpha_1} 3^{\alpha_2} 5^{\alpha_3} 7^{\alpha_4} 11^{\alpha_5} \cdot m$, where m is a positive integer which cannot be divided by 2, 3, 5, 7, 11, α_1, α_2, α_3, α_4, α_5 are natural numbers. Choose S to contain the positive integers such that there are only 1 or 2 positive integers in α_1, α_2, α_3, α_4, α_5; that is to say, S includes the 50 even numbers 2, 4, $\ldots$, 100 except $2 \times 3 \times 5$, $2^2 \times 3 \times 5$, $2 \times 3^2 \times 5$, $2 \times 3 \times 7$, $2^2 \times 3 \times 7$, $2 \times 5 \times 7$, $2 \times 3 \times 11$, odd multiples of 3(3×1, 3×3, $\ldots$, 3×33, which are 11 numbers in total), odd numbers whose minimal prime divisor is 5 (this includes the seven numbers 5×1, 5×5, 5×7, 5×11, 5×13, 5×17, 5×19), odd numbers whose minimal prime divisor is 7 (this includes the four numbers 7×1, 7×7, 7×11, 7×13), and the prime number 11. Then $|S| = (50 - 7) + 17 + 7 + 4 + 1 = 72$. Now we prove that S satisfies the requirements.

Condition (1) is clearly satisfied.

Consider the condition (2). For any two different elements a, b in S, the prime divisors of $[a, b]$ can include at most four numbers in 2, 3, 5, 7, 11; assume that p is not included, then $p \in S$, and $(p, a) \leqslant (p, [a, b]) = 1$, $(p, b) \leqslant (p, [a, b]) = 1$. We let $c = p$, and see that the condition is satisfied.

Consider the condition (3). When $(a, b) = 1$, choose the minimum prime factor p of a and the minimum prime factor q of b, then apparently $p \neq q$, and p, $q \in \{2, 3, 5, 7, 11\}$, thus $pq \in S$, and $(pq, a) \geqslant p > 1$, $(pq, b) \geqslant q > 1$. Since a, b are relatively prime, pq is different from a, b. Then let $c = pq$ and see that the condition is satisfied. When $(a, b) = e > 1$, choose the minimum prime factor p of a, and the minimum prime number q which does divide $[a, b]$, then apparently $p \neq q$, and p, $q \in \{2, 3, 5, 7, 11\}$, so $pq \in S$, and $(pq, a) \geqslant (p, a) = p > 1$, $(pq, b) \geqslant (p, b) = p > 1$. Since q cannot divide $[a, b]$, pq is different from a, b. So we can choose $d = pq$.

Next we prove that, for any such set S, $|S| \leqslant 72$.

Apparently, $1 \notin S$. For any prime numbers p, q which are larger than 10, since the minimal integer which is not relatively prime to p and q is pq, and $pq > 100$, from condition (3), we know that at most one of the 21 prime numbers 11, 13, $\ldots$, 89, 97 can belong to S. Denote the set including the 78 positive numbers which are no more than 100 and different from 1 and these 21 prime numbers by T. We prove that there are at least seven numbers in T that are not in S, thus $|S| \leqslant 78 - 7 + 1 = 72$.

Actually, if some prime number $p \in S$ is larger than 10, then the minimal prime divisor of the numbers in S can only be one of 2, 3, 5, 7 and p. From condition (2), we know that:

(i) if $7p \in S$, since $2 \times 3 \times 5$, $2^2 \times 3 \times 5$, $2 \times 3^2 \times 5$ and $7p$ includes all the possible minimal prime divisors, then by condition (2), $2 \times 3 \times 5$, $2^2 \times 3 \times 5$, $2 \times 3^2 \times 5 \notin S$. If $7p \notin S$, and $2 \times 7p > 100$, then $p \in S$. Thus by condition (3), 7×1, 7×7, 7×11, $7 \times 13 \notin S$.

(ii) If $5p \in S$, then $2 \times 3 \times 7$, $2^2 \times 3 \times 7 \notin S$. If $5p \notin S$, then 5×1, $5 \times 5 \notin S$.

(iii) $2 \times 5 \times 7$ and $3p$ do not simultaneously belong to S.

(iv) $2 \times 3p$ and 5×7 do not simultaneously belong to S.

(v) If $5p$, $7p \notin S$, then $5 \times 7 \notin S$.

Therefore, when $p = 11$ or 13, from (i)–(iv) we can see that there are at least 3, 2, 1, 1 numbers in T not belonging to S, which are seven numbers in total. If $p = 17$ or 19, by (i)–(iii) we can see there are at least 4, 2, 1 numbers in T not belonging to S, and there are seven numbers in total. When $p > 20$, by (i)–(iii) we can see there are at least four, two, one numbers in T not belonging to S, and there are seven numbers in total. So the conclusion holds.

When there is no prime number larger than 10 belonging to S, the minimum prime divisor of the numbers in S can only be 2, 3, 5, 7. Thus, each of the seven sets $\{3, 2 \times 5 \times 7\}$, $\{5, 2 \times 3 \times 7\}$, $\{7, 2 \times 3 \times 5\}$, $\{2 \times 3, 5 \times 7\}$, $\{2 \times 5, 3 \times 7\}$, $\{2 \times 7, 3 \times 7\}$, $\{2^2 \times 7, 3^2 \times 5\}$ should contain at least one element which does not belong to S. The conclusion holds.

In conclusion, $|S|_{\max} = 72$.

Example 7. Let a_i, b_i ($i = 1, 2, \ldots, n$) be rational numbers, such that for any real number x, we have $x^2 + x + 4 = \sum_{i=1}^{n} (a_i x + b_i)^2$, find the minimum value of n. (2006 Chinese National Team)

Solution 1. It is easy to see that $n = 5$ is possible, actually

$$x^2 + x + 4 = \left(x + \frac{1}{2}\right)^2 + \left(\frac{3}{2}\right)^2 + 1^2 + \left(\frac{1}{2}\right)^2 + \left(\frac{1}{2}\right)^2.$$

Then we guess that the minimum value of n is 5, and we only have to prove that $n \neq 4$.

Assume the contrary that $x^2 + x + 4 = \sum_{i=1}^{4} (a_i x + b_i)^2$, a_i, $b_i \in \mathbf{Q}$, then $\sum_{i=1}^{4} a_i^2 = 1$, $\sum_{i=1}^{4} a_i b_i = \frac{1}{2}$, $\sum_{i=1}^{4} b_i^2 = 4$.

Then,

$$\frac{15}{4} = \left(\sum_{i=1}^{4} a_i^2\right)\left(\sum_{i=1}^{4} b_i^2\right) - \left(\sum_{i=1}^{4} a_i b_i\right)^2$$

$$= (-a_1 b_2 + a_2 b_1 - a_3 b_4 + a_4 b_3)^2 + (-a_1 b_3 + a_3 b_1 - a_4 b_2 + a_2 b_4)^2$$
$$+ (-a_1 b_4 + a_4 b_1 - a_2 b_3 + a_3 b_2)^2.$$

Multiplying by 4 both side of the equation, we know that $a^2 + b^2 + c^2 = 15d^2 \equiv -d^2 \pmod 8$ has a solution.

We may assume that there is at least one odd number among a, b, c, d (otherwise divide both sides by the common divisor), but

$$a^2, b^2, c^2, d^2 \equiv 0, 1, 4 \pmod 8,$$

so the above equation has no solution, which is a contradiction.

Solution 2. When $n = 5$, we have

$$x^2 + x + 4 = \left(\frac{1}{2}x + 1\right)^2 + \left(\frac{1}{2}x + 1\right)^2 + \left(\frac{1}{2}x - 1\right)^2 + \left(\frac{1}{2}x\right)^2 + (1)^2.$$

Next we prove that $n \neq 4$.

If $x^2 + x + 4 = \sum_{i=1}^{4} (a_i x + b_i)^2$, we can assume $a_i = \frac{x_i}{2m}$, $b_i = \frac{y_i}{k}$ $(mk \neq 0, m, k, x_i, y_i \in \mathbf{Z})$, then by comparing the coefficients we get that

$$\begin{cases} x_1^2 + x_2^2 + x_3^2 + x_4^2 = 4m^2, & \text{①} \\ y_1^2 + y_2^2 + y_3^2 + y_4^2 = 4n^2, & \text{②} \\ x_1 y_1 + x_2 y_2 + x_3 y_3 + x_4 y_4 = mn. & \text{③} \end{cases}$$

Assume that (m, k, x_i, y_i) is the minimal solution which satisfies ①
②③ and makes $\mid mk \mid$ nonzero. From ①, $x_1^2 + x_2^2 + x_3^2 + x_4^2 \equiv 0 \pmod 4$.
Since $x^2 \equiv 0$ or $1 \pmod 4$ we can see that all of the x_i's have the same
parity. Similarly, all of the y_i's have the same parity.

Therefore, $x_i y_i (i = 1, 2, 3, 4)$ have the same parity. From ③,
mn is even, assume m is even, then by ①,

$$x_1^2 + x_2^2 + x_3^2 + x_4^2 \equiv 0 \pmod 8.$$

If all of the x_i are odd, then $x_i^2 \equiv 1 \pmod 8$, thus $x_1^2 + x_2^2 + x_3^2 + x_4^2 \equiv 4 \pmod 8$, which is a contradiction.

Therefore all of the x_i's are even.

Thus $\left(\dfrac{m}{2}, k, \dfrac{x_i}{2}, y_i \right)$ also satisfies ①②③ and makes $\mid mk \mid$ nonzero,
and is smaller, which is a contradiction.

Example 8. If the number of elements in a set S is even, then the set is
called an even set. Let $M = \{1, 2, \ldots, 2012\}$. If there are k even
subsets $A_1, A_2, \ldots, A_k$ of M, such that for any $1 \leqslant i < j \leqslant k$, $A_i \cap A_j$ is not a even set, what is the maximum value of k? (Original)
Solution. First, it is easy to see that $k = 2011$ satisfies the requirement.
Actually, let $A_i = \{i, 2012\} (i = 1, 2, \ldots, 2011)$, then for any $1 \leqslant i < j \leqslant 2011$, we have $A_i \cap A_j = \{2012\}$, which is not an even set.
Thus $k = 2011$ satisfies the requirements.

Next we will prove that $k \leqslant 2011$. We proceed by assuming $k \geqslant 2012$, thus there are 2012 even subsets of M, say $A_1, A_2, \ldots, A_{2012}$, such that for any $1 \leqslant i < j \leqslant 2012$, $A_i \cap A_j$ is not an even set.

For $A \subseteq M$, let $\vec{\alpha}_A = \{a_1, a_2, \ldots, a_{2012}\}$, where

$$a_j = \begin{cases} 1, & \text{if } j \in A, \\ 0, & \text{if } j \notin A. \end{cases}$$

We know that $A_i \cap A_j$ is a even set if and only if $\overrightarrow{\alpha A_i} \cdot \overrightarrow{\alpha A_j}$ is even; thus for any $i \neq j$, we have

$$\overrightarrow{\alpha A_i} \cdot \overrightarrow{\alpha A_j} \equiv 1 \pmod 2. \qquad \textcircled{1}$$

For $X \subseteq M$, let $\overrightarrow{S_X} = \sum_{x \in X} \overrightarrow{\alpha A_x}$, we first prove that, for any $X \neq \varnothing$, we have $\overrightarrow{S_X} \not\equiv (0, 0, \ldots, 0) \pmod 2$.

Actually, suppose $\overrightarrow{S_X} \equiv (0, 0, \ldots, 0)$. On the one hand, let $u \in X$, we have

$$0 \equiv \overrightarrow{S_X} \cdot \overrightarrow{\alpha A_u} = \left(\sum_{x \in X} \overrightarrow{\alpha A_x} \right) \cdot \overrightarrow{\alpha A_u} = \sum_{x \in X} \left(\overrightarrow{\alpha A_x} \cdot \overrightarrow{\alpha A_u} \right)$$
$$= \overrightarrow{\alpha A_u} \cdot \overrightarrow{\alpha A_u} + \sum_{x \in X \setminus \{u\}} \left(\overrightarrow{\alpha A_x} \cdot \overrightarrow{\alpha A_u} \right).$$

Since A_u is an even set, we have $\overrightarrow{\alpha A_u} \cdot \overrightarrow{\alpha A_u} \equiv 0$. Also when $x \neq u$, from $\textcircled{1}$ we have $\overrightarrow{\alpha A_x} \cdot \overrightarrow{\alpha A_u} \equiv 1$. Thus

$$0 \equiv \overrightarrow{\alpha A_u} \cdot \overrightarrow{\alpha A_u} + \sum_{x \in X \setminus \{u\}} \left(\overrightarrow{\alpha A_x} \cdot \overrightarrow{\alpha A_u} \right) \equiv 0 + \sum_{x \in X \setminus \{u\}} 1 = |X| - 1 \pmod 2,$$

thus $|X|$ is odd.

On the other hand, note that $|M| = 2012$ is even, we know $X \neq M$. Choosing $v \notin X$, we have

$$0 \equiv \overrightarrow{S_X} \cdot \overrightarrow{\alpha A_v} = \left(\sum_{x \in X} \overrightarrow{\alpha A_x} \right) \cdot \overrightarrow{\alpha A_v} = \sum_{x \in X} \left(\overrightarrow{\alpha A_x} \cdot \overrightarrow{\alpha A_v} \right).$$

Since $v \notin X$, we know that $x \neq v$, then from $\textcircled{1}$, $\overrightarrow{\alpha A_x} \cdot \overrightarrow{\alpha A_v} \equiv 1$. Therefore

$$0 \equiv \sum_{x \in X} \left(\overrightarrow{\alpha A_x} \cdot \overrightarrow{\alpha A_v} \right) \equiv \sum_{x \in X} 1 = |X|,$$

so $|X|$ is even, which is a contradiction.

Therefore, for any $X \neq \varnothing$, we have $\overrightarrow{S_X} \not\equiv (0, 0, \ldots, 0) \pmod 2$.

Therefore, for $X \neq Y$, we cannot have $\overrightarrow{S_X} \equiv \overrightarrow{S_Y} \pmod 2$ (in fact, if $\overrightarrow{S_X} \equiv \overrightarrow{S_Y}$, let $T = (X \cup Y) \setminus (X \cap Y)$, then we have $\overrightarrow{S_T} = \overrightarrow{S_X} + \overrightarrow{S_Y} - 2 \overrightarrow{S_{X \cap Y}} \equiv \overrightarrow{S_X} + \overrightarrow{S_Y} \equiv (0, 0, \ldots, 0)$, which is a contradiction).

Thus X runs over all subsets of M, we get 2^{2012} different vectors $\overrightarrow{S_X}$ when mod 2.

However, since $\overrightarrow{S_X} = \sum_{x \in X} \overrightarrow{\alpha A_x}$ and A_x is an even set, we know that the sum of the entries of A_x is even. Thus the parity of the 2012th entry of $\overrightarrow{S_X}$ is determined by the parity of the sum of the previous 2011 entries. Thus $\overrightarrow{S_X}$ can have at most 2^{2011} choices mod 2, which is a contradiction. So we know $k \leqslant 2011$.

In conclusion, the maximum value of k is 2011.

Note. Apparently, 2012 can be replaced by any positive even integer n, and the corresponding maximum value of k is $n - 1$.

Exercise 8

(1) The natural number n satisfies the following conditions: choose n different odd numbers in 1, 2, $\ldots$, 100, there must be two numbers whose sum is 102. Find the minimum value of n.

(2) Let $X = \{1, 2, \ldots, 1995\}$, and A be a subset of X. If for any two elements x, $y(x < y)$ in A, there is $y \neq 15x$, find the maximum value of $|A|$.

(3) Let $X = \{0, 1, 2, \ldots, 9\}$, $F = \{A_1, A_2, \ldots, A_k\}$ where each element A_i is a nonempty subset of X, and for any $1 \leqslant i < j \leqslant k$, there is $|A_i \cap A_j| \leqslant 2$. Find the maximum value of k. (26th IMO alternative)

(4) Let

$$A_i = \{i, i+1, i+2, \ldots, i+59\} \ (i = 1, 2, \ldots, 11),$$
$$A_{11+j} = \{11+j, 12+j, \ldots, 70, 1, 2, \ldots, j\} \ (j = 1, 2, \ldots, 59).$$

There are k sets in these 70 sets, where the intersection of any 7 of them is nonempty. Find the maximum value of k.

(5) Let S is a nonempty subset of set $\{1, 2, \ldots, 108\}$, satisfying:
(i) for any number a, b in S, there is always a number c, such that $(a, c) = (b, c) = 1$; (ii) for any number a, b in S, there is always a

number c' in S, such that $(a, c') > 1$, $(b, c') > 1$. Find the maximum value of the number of elements in S. (2004 Chinese National Team)

(6) Find the minimum positive integer n with the following property: paint the vertices of a regular n-polygon to any of the three colors red, blue, yellow. Then there must be four vertices with the same color, which form a isosceles trapezoid. (2008 China Mathematical Olympiad)

Chapter 9   Global Estimates

To estimate the range of some variable, we may put it in a set of several variables and consider them together, and obtain some global bounds. Using this we can obtain some information on the range of the variable. This method is called global estimation.

A special case of global estimates is to estimate the mean value: assume that the mean of A_1, A_2,..., A_n is A, then at least one number in A_1, A_2, ... , A_n is not less than A, and at least one number in A_1, A_2, ... , A_n is no more than A.

One of the important tools used in global estimates is the "*relation table of the elements and subsets*": assume that $X = \{a_1, a_2, \ldots, a_n\}$, and A_1, A_2, ... ,A_k are subsets of X. The so-called "*relation table of the elements and subsets*" means the following table with n rows and k columns:

elements＼Subsets F	A_1	A_2	$\cdots$	A_k	
a_1	x_{11}	x_{12}	$\cdots$	x_{1k}	m_1 1s
a_2	x_{21}	x_{22}	$\cdots$	x_{2k}	m_2 1s
$\vdots$	$\vdots$				$\vdots$
a_n	x_{n1}	x_{n2}	$\cdots$	x_{nk}	m_n 1s

Here when $a_i \in A_j$, we have $x_{ij} = 1$, otherwise, $x_{ij} = 0$. The number of 1s in the ith row means the number of appearance of element a_i in the subsets, which is called the degree of a_i, and denoted as $d(a_i)$ or m_i, thus $m_i = \sum_{j=1}^{k} x_{ij}$.

The number of $1s$ in the j th column is the number of elements in set A_j, which is $|A_j| = \sum_{i=1}^{n} x_{ij}$.

There are two useful formulas in this table:

(1) Consider the total number of appearances of the elements in F, that is, the number of $1s$ in the table, we have

$$\sum_{i=1}^{n} m_i = S = \sum_{j=1}^{k} |A_j|.$$

(2) Consider the total number T of appearances of elements in the intersections of two subsets, we have

$$\sum_{i=1}^{n} C_{m_i}^2 = T = \sum_{1 \leqslant i < j \leqslant k} |A_i \cap A_j|.$$

Example 1. Find the largest positive integer A, such that for any permutation of $1, 2, \ldots, 100$, there are 10 consecutive terms whose sum is no less than A. (22nd Poland Mathematical Olympiad)

Analysis. For a permutation of $1, 2, \ldots, 100$, it is difficult to find 10 consecutive terms whose sum is no less than A. We can consider the sums of all of the consecutive terms which share no common terms with each other, and estimate the mean value.

Solution. Let $a_1, a_2, \ldots, a_{100}$ be a permutation of $1, 2, \ldots, 100$. Let $A_i = a_i + a_{i+1} + \cdots + a_{i+9} (i = 1, 2, \ldots, 91)$; then $A_1 = a_1 + a_2 + \cdots + a_{10}$, $A_{11} = a_{11} + a_{12} + \cdots + a_{20}, \ldots, A_{91} = a_{91} + a_{92} + \cdots + a_{100}$.

Note that $A_1 + A_{11} + \cdots + A_{91} = a_1 + a_2 + \cdots + a_{100} = 5050$. Thus the mean of $A_1, A_{11}, \ldots, A_{91}$ is 505, thus there is at least one $i (1 \leqslant i \leqslant 91)$, such that $A_i \geqslant 505$.

When $A \geqslant 506$, consider the permutation: $(100, 1, 99, 2, 98, 3, 97, 4, \ldots, 51, 50)$, where the terms are $a_{2i-1} = 101 - i$ $(i = 1, 2, \ldots, 50)$; $a_{2i} = i (i = 1, 2, \ldots, 50)$. So $a_{2i-1} + a_{2i} = 101$, $a_{2i} + a_{2i+1} = 100$. Now we can prove that for any $i (1 \leqslant i \leqslant 91)$, $A_i \leqslant 505 < A$.

Actually, when i is even, let $i = 2k$.

Then

$$A_{2k} = a_{2k} + a_{2k+1} + \cdots + a_{2k+9}$$
$$= (a_{2k} + a_{2k+1}) + (a_{2k+2} + a_{2k+3}) + \cdots + (a_{2k+8} + a_{2k+9})$$
$$= 100 \times 5$$
$$= 500 < A.$$

When i is odd, let $i = 2k - 1$. Then

$$A_{2k-1} = a_{2k-1} + a_{2k} + \cdots + a_{2k+8}$$
$$= (a_{2k-1} + a_{2k}) + (a_{2k+1} + a_{2k+2}) + \cdots + (a_{2k+7} + a_{2k+8})$$
$$= 101 \times 5$$
$$= 505 < A.$$

So $A_{\max} = 505$.

Example 2. Let $A_i (i = 1, 2, \ldots, 30)$ be subsets of $M = \{1, 2, 3, \ldots, 1990\}$, $|A_i| \geqslant 660$. Prove that: there exist $i, j (1 \leqslant i < j \leqslant 30)$, such that $|A_i \cap A_j| \geqslant 200$.

Proof. Assume that all $|A_i| = 660$. Otherwise, remove some elements in A_i and obtain A_i'. If we can prove that $|A_i' \cap A_j'| \geqslant 200$, then adding the removed elements, we certainly have $|A_i \cap A_j| \geqslant 200$.

Consider the relation table as above. Assume that there are m_i 1s in the ith column; that is, i appears in m_i subsets. Consider the total number of appearances of each element in all subsets, or the number of 1s in the table, we have

$$\sum_{i=1}^{1990} m_i = S = \sum_{i=1}^{30} |A_i| = 30 \times 660.$$

Now estimate $\sum_{1 \leqslant i < j \leqslant 30} |A_i \cap A_j|$. This is just the total number of appearance of elements in intersections of subsets, thus we have

$$\sum_{i=1}^{1990} C_{m_i}^2 = \sum_{1 \leqslant i < j \leqslant 30} |A_i \cap A_j|.$$

Then, from the Cauchy-Schwartz inequality, we have

$$2 \sum_{1 \leqslant i < j \leqslant 30} |A_i \cap A_j| = 2 \sum_{i=1}^{1990} C_{m_i}^2 = \sum_{i=1}^{1990} m_i^2 - \sum_{i=1}^{1990} m_i$$

$$\geq \frac{\left(\sum\limits_{i=1}^{1990} m_i\right)^2}{\sum\limits_{i=1}^{1990} 1^2} - \sum\limits_{i=1}^{1990} m_i$$

$$= \frac{(30 \times 660)^2}{1990} - 30 \times 660.$$

So there must exist (i, j) such that

$$|A_i \cap A_j| \geq \frac{\dfrac{(30 \times 660)^2}{1990} - 30 \times 660}{2C_{30}^2}$$

$$= \frac{(30 \times 660) \times (30 \times 660 - 1990)}{30 \times 29 \times 1990}$$

$$> 200.$$

Example 3. Ten people go to a bookstore to buy books. We know that each person has bought three books, and that for any two people, there is at least one book bought by both of them. Consider the book that is the most popular (that is, bought by the largest number of people). At least how many people bought this book? (8th China Mathematical Olympiad)

Solution. Assume that they have bought n books, and the set of books bought by the ith person is $A_i(i = 1, 2, \ldots, 10)$. Construct the relation table, and assume that the ith row has m_i 1s.

Counting the number of appearances of elements, we have

$$\sum_{i=1}^{n} m_i = S = \sum_{i=1}^{10} |A_i| = \sum_{i=1}^{10} 3 = 30.$$

Then we count the number of appearances of elements in the intersections of subsets, and get

$$\sum_{i=1}^{n} C_{m_i}^2 = \sum_{1 \leqslant i < j \leqslant 10} |A_i \cap A_j|.$$

Assume the maximum one of m_i is m, then

$$90 = 2C_{10}^2 = 2\sum_{1 \le i < j \le 10} 1 \le 2\sum_{1 \le i < j \le 10} |A_i \cap A_j|$$

$$= 2\sum_{i=1}^n C_{m_i}^2 = \sum_{i=1}^n m_i^2 - \sum_{i=1}^n m_i = \sum_{i=1}^n m_i^2 - 30$$

$$\le \sum_{i=1}^n (m_i \cdot m) - 30 = m\sum_{i=1}^n m_i - 30$$

$$= 30(m - 1). \qquad \qquad ①$$

Solving this inequality, we have $m \ge 4$. If $m = 4$, then equality holds in ①, so all $m_i = 4(i = 1, 2, \ldots, n)$. Thus we have $4n = \sum_{i=1}^n m_i = 30$, which is a contradiction. Thus $m \ge 5$.

When $m = 5$, since m is the maximum one, all of $m_i \le 5$. Thus $30 = \sum_{i=1}^n m_i \le 5n$, $n \ge 6$. We choose $n = 6$, and a possible construction is shown below:

	A_1	A_2	A_3	A_4	A_5	A_6	A_7	A_8	A_9	A_{10}
1	*	*	*	*	*					
2	*	*				*	*	*		
3	*		*			*			*	*
4				*	*	*	*		*	
5			*	*			*	*		*
6		*			*			*	*	*

So the minimum value of m is 5.

Example 4. A group of boy scouts aged from 7 to 13 (ages are integers) are coming from 11 countries. Prove that there are at least five children such that for any one of them, there are more people having the same age with him than those having the same nationality. (Canada National Team)

Proof. Consider the weighted relation table:

	A_1	A_2	...	A_{11}
7	$a_{1,1}$	$a_{1,2}$	...	$a_{1,11}$
8	$a_{2,1}$	$a_{2,2}$	...	$a_{2,11}$
$\vdots$	$\vdots$	$\vdots$	$\vdots$	$\vdots$
13	$a_{13,1}$	$a_{13,2}$	...	$a_{13,11}$

where A_j is the set of people coming from the jth country, and a_{ij} is the number of people with age I in country j. Assume that the sum of the ith row is r_i and the sum of the jth column is t_j, then

$$\sum_{i=7}^{13}\sum_{j=1}^{11} a_{ij}\left(\frac{1}{t_j}-\frac{1}{r_i}\right)$$

$$=\sum_{i=7}^{13}\sum_{j=1}^{11}\frac{a_{ij}}{t_j}-\sum_{i=7}^{13}\sum_{j=1}^{11}\frac{a_{ij}}{r_i}$$

$$=\sum_{j=1}^{11}\frac{1}{t_j}\sum_{i=7}^{13}a_{ij}-\sum_{i=7}^{13}\frac{1}{r_i}\sum_{j=1}^{11}a_{ij}$$

$$=\sum_{j=1}^{11}\left(\frac{1}{t_j}\cdot t_j\right)-\sum_{i=7}^{13}\left(\frac{1}{r_i}\cdot r_i\right)$$

$$=\sum_{j=1}^{11}1-\sum_{i=7}^{13}1=4.$$

Since $\dfrac{1}{t_j}-\dfrac{1}{r_i}<1$, considering the term $a_{ij}\left(\dfrac{1}{t_j}-\dfrac{1}{r_i}\right)$ in the above expression as the sum of a_{ij} "$\dfrac{1}{t_j}-\dfrac{1}{r_i}$", we know that there are at least five such terms which are positive in the above expression. Thus there are five children who satisfy the requirement.

Example 5. Let $A=\{1,2,3,4,5,6\}$, $B=\{7,8,9,\ldots,n\}$. For each i, choose three numbers in A and two numbers in B, and form a set $A_i(i=1,2,\ldots,20)$ with five elements, such that $|A_i\cap A_j|\leqslant 2$, $1\leqslant i<j\leqslant 20$, find the minimum value of n. (2002 IMO China National Team Selection)

Analysis. This question actually requires to find the minimum value of $|B|$. The total appearances of elements in B in the subsets $A_i(i=1,$

2, ... , 20) is $2 \times 20 = 40$. To find out how many elements are there in B, we only have to know at most how many times each element in B can appear in the subsets A_i $(i = 1, 2, ... , 20)$.

Solution. We first prove that each element in B can appear at most four times in the subsets A_i $(i = 1, 2, ... , 20)$. Otherwise, we assume that some element b in B appears k $(k > 4)$ times in A_i $(i = 1, 2, ... , 20)$. Consider the k subsets containing b, they have in total $3k > 12$ elements in A. Then by the pigeonhole principle, there is at least one element in A (assume it is a), which appears three times in the k subsets. Assume that the three subsets containing a, b are P, Q, R, then the five elements in $A \backslash \{a\}$ appear six times in P, Q, R. So there must be one element c which appears twice. Then we have two sets both containing a, b, c, which contradicts the condition $| A_i \cap A_j | \leqslant 2$.

By the above discussion, each element in B appears in the subsets A_i $(i = 1, 2, ... , 20)$ at most four times, and the total appearances of elements in B in each subset A_i $(i = 1, 2, ... , 20)$ is 40, so $| B | \geqslant \dfrac{40}{4} = 10$, so $n \geqslant 10 + 6 = 16$.

Finally, when $n = 16$, we can construct the 20 sets satisfying the requirements: $\{1, 2, 3, 7, 8\}$, $\{1, 2, 4, 12, 14\}$, $\{1, 2, 5, 15, 16\}$, $\{1, 2, 6, 9, 10\}$, $\{1, 3, 4, 10, 11\}$, $\{1, 3, 5, 13, 14\}$, $\{1, 3, 6, 12, 15\}$, $\{1, 4, 5, 7, 9\}$, $\{1, 4, 6, 13, 16\}$, $\{1, 5, 6, 8, 11\}$, $\{2, 3, 4, 13, 15\}$, $\{2, 3, 5, 9, 11\}$, $\{2, 3, 6, 14, 16\}$, $\{2, 4, 5, 8, 10\}$, $\{2, 4, 6, 7, 11\}$, $\{2, 5, 6, 12, 13\}$, $\{3, 4, 5, 12, 16\}$, $\{3, 4, 6, 8, 9\}$, $\{3, 5, 6, 7, 10\}$, $\{4, 5, 6, 14, 15\}$.

In conclusion, the minimum value of n is 16.

Example 6. Let n be a given positive integer with $6 \mid n$. In an $n \times n$ chessboard, each cell is filled with a positive integer. The numbers filled in the ith row are $(i - 1)n + 1$, $(i - 1)n + 2$, ... , $(i - 1)n + n$ from left to right. In each step we choose two neighboring (sharing a common side) cells, and add 1 to one number and add 2 to the other number. How many steps does it need to make the numbers filled in each cells equal? (Original)

Solution. Denote the sum of the numbers in the chessboard by S.

Apparently, when the numbers are equal, each number should be at least n^2. Thus the sum of the numbers in the chessboard is given by $S' \geqslant n^2 \cdot n^2 = n^4$. Since each operation increases S by 3, and in the beginning $S_0 = 1 + 2 + 3 + \cdots + n^2 = \dfrac{n^2(n^2+1)}{2}$, so when the numbers are equal, S has increased by at least $n^4 - \dfrac{n^2(n^2+1)}{2} = \dfrac{n^2(n^2-1)}{2}$. Thus the number of operations is no less than $\dfrac{1}{3} \cdot \dfrac{n^2(n^2-1)}{2} = \dfrac{n^2(n^2-1)}{6}$.

Next we prove that, we can use $\dfrac{n^2(n^2-1)}{6}$ steps to equalize the numbers filled in the cells. It is equivalent to prove that: we can use the operation to make every number n^2.

In each row, group every three consecutive cells together, and divide the row to $\dfrac{n}{3}$ groups, which are marked as group 1, group 2, ..., group $\dfrac{n}{3}$.

First we notice that for any three consecutive cells, we use two steps to increase each number by 2. We call such two steps a big step A: $(a, b, c) \rightarrow (a+2, b+1, c) \rightarrow (a+2, b+2, c+2)$. Now we do $\dfrac{n}{2}$ big step A's in each group of some row, so that the numbers in this row will increase by n. Then, for each row, we can do several such operations, so that this row becomes the same as the nth row.

Now note that for any two consecutive cells, we can use two steps to increase each number by 3. We call such two operations a big step B: $(a, b) \rightarrow (a+2, b+1) \rightarrow (a+3, b+3)$. Now, assume that each row of the chessboard is the same as the nth row. Consider the jth group in the ith row and the $(i+1)$th row, and use an operation B to the two cells in the same column of these two groups, so that the numbers in the jth group of the two rows increase by 3. Since the differences between the corresponding numbers in different groups in the same row are multiples of 3, we can do several operations B to the two rows, so that the all of the groups in the two rows become the same as the group $\frac{n}{3}$, which is $(n^2 - 2, n^2 - 1, n^2)$. After one operation, it can be changed to (n^2, n^2, n^2), so the two rows can be changed to n^2.

We then repeat the operations, until each cell becomes n^2.

In conclusion, the minimum number of operations is $\dfrac{n^2(n^2 - 1)}{6}$.

Example 7. Given a positive integer n, let $X = \{1, 2, 3, \ldots, n\}$, and A be a subset of X. For any $x < y < z$, x, y, $z \in A$, there is a triangle whose side lengths are x, y, z respectively. Denote by $|A|$ the number of elements in A, find the maximum value of $|A|$. (Original)

Solution. Let $A = \{a_1, a_2, \ldots, a_r\}$ be such a subset, where $a_1 < a_2 < \cdots < a_r$. If $a_r < n$, let $n - a_r = t$, and $a'_i = a_i + t (i = 1, 2, \ldots, r)$, then $a'_r = n$, and $A' = \{a'_1, a'_2, \ldots, a'_r\}$ also satisfies the requirement. Thus, we can assume $a_r = n$.

Since a_1, a_2, a_r are the side lengths of a triangle, we have $n = a_r < a_1 + a_2$. Since $a_1 + a_2 < 2a_2$, we have $a_1 + a_2 \leqslant 2a_2 - 1$, thus $n = a_r \leqslant 2a_2 - 2$, so $a_2 \geqslant \frac{1}{2}n + 1$.

We now consider what numbers can be chosen in $A \backslash \{a_1\}$.

(1) If n is odd, let $n = 2k + 1$, then $a_2 \geqslant \frac{1}{2}n + 1 = k + \frac{3}{2}$, but a_2 is integer, so $a_2 \geqslant k + 2$. Thus, $A \backslash \{a_1\} \subseteq \{k + 2, k + 3, \ldots, 2k + 1\}$, thus $|A| - 1 \leqslant |\{k + 2, k + 3, \ldots, 2k + 1\}| = 2k + 1 - (k + 1) = k$, so $|A| \leqslant k + 1 = \dfrac{n - 1}{2} + 1 = \dfrac{n + 1}{2}$.

(2) If n is even, let $n = 2k$, then $a_2 \geqslant \frac{1}{2}n + 1 = k + 1$. Thus,

$A\backslash\{a_1\} \subseteq \{k+1, k+2, \ldots, 2k\}$, so $|A|-1 \leqslant |\{k+1, k+2, \ldots,$
$2k\}| = 2k - k = k$, so $|A| \leqslant k+1 = \frac{n}{2}+1 = \frac{n+2}{2}$.

From (1), (2), for any positive integer n, we have $|A| \leqslant \frac{n+2}{2}$.

Since $|A|$ is a positive integer, we have $|A| \leqslant \left[\frac{n+2}{2}\right]$.

Next, if $n = 2k$, let $A = \{k, k+1, k+2, \ldots, 2k\}$, then for any
$x < y < z$, $x, y, z \in A$, we have $x+y \geqslant k+(k+1) = 2k+1 > 2k \geqslant$
z, thus there exists a triangle with x, y, z as its three side lengths,
and $|A| = k+1 = \frac{n}{2}+1 = \frac{n+2}{2} = \left[\frac{n+2}{2}\right]$.

If $n = 2k+1$, then let $A = \{k+1, k+2, \ldots, 2k+1\}$. For any
$x < y < z$, $x, y, z \in A$, we have $x+y \geqslant (k+1)+(k+2) = 2k+$
$3 > 2k+1 \geqslant z$, thus there exists a triangle with x, y, z as its three side
lengths, and $|A| = k+1 = \frac{n-1}{2}+1 = \frac{n+1}{2} = \left[\frac{n+2}{2}\right]$. $\left(\text{Note: the}\right.$
constructions in the above two scenarios can be merged into one: $A =$
$\left.\left\{\left[\frac{n+1}{2}\right], \left[\frac{n+1}{2}\right]+1, \left[\frac{n+1}{2}\right]+2, \ldots, n\right\}.\right)$

Example 8. Given a positive integer $n(n \geqslant 2)$, find the max λ, such
that: if there are n bags, each bag containing some balls whose weights
are integer powers of 2, and the total weights in each bag equal each
other (there can be balls with the same weight in the same bag), then
there must exists λ balls with the same weight. (2005 China National
Team Selection)
Solution. Assume that the heaviest ball has weight 1, and the total
weight of balls in each bag is G, then $G \geqslant 1$.

First we prove that: $\lambda = \left[\frac{n}{2}\right]+1$ satisfies the condition, that is,

there must exist at least $\left[\frac{n}{2}\right]+1$ balls having the same weight.

We proceed by contradiction. Assume that the total number of balls with any weight is no more than $\left[\dfrac{n}{2}\right]$. Considering the total weight of the n bags, we have

$$n \leqslant nG < \left[\frac{n}{2}\right] \cdot (1 + 2^{-1} + 2^{-2} + \cdots) = 2\left[\frac{n}{2}\right] \leqslant 2 \cdot \frac{n}{2} = n,$$

which is a contradiction.

Next we prove: $\lambda \leqslant \left[\dfrac{n}{2}\right] + 1$. Find a positive integer s which is sufficiently large, such that $2 - 2^{-s} \geqslant \dfrac{2n}{n+1}$, since $\left[\dfrac{n}{2}\right] + 1 \geqslant \dfrac{n+1}{2}$, we know

$$2 - 2^{-s} \geqslant \frac{2n}{n+1} \geqslant \frac{n}{\left[\frac{n}{2}\right] + 1},$$

thus $\left(\left[\dfrac{n}{2}\right] + 1\right)(1 + 2^{-1} + \cdots + 2^{-s}) \geqslant n \cdot 1$.

So from

$$\underbrace{1, 1, \ldots, 1}_{\left[\frac{n}{2}\right]+1}, \underbrace{2^{-1}, 2^{-1}, \ldots, 2^{-1}}_{\left[\frac{n}{2}\right]+1}, \ldots, \underbrace{2^{-s}, 2^{-s}, \ldots, 2^{-s}}_{\left[\frac{n}{2}\right]+1},$$

we can extract at least n sets of consecutive terms whose sum is 1, so $\lambda \leqslant \left[\dfrac{n}{2}\right] + 1$.

In conclusion, $\lambda_{\max} = \left[\dfrac{n}{2}\right] + 1$.

Exercise 9

(1) Assume that $A_1, A_2, \ldots, A_{29}$ are 29 different sets of positive integers. For $1 \leqslant i < j \leqslant 29$ and positive integer x, define $N_i(x) = $ the number of elements in A_i which is no more than x, and $N_{ij}(x) = $ the number of elements in $A_i \cap A_j$ which is no more than x. Given that

for all $i = 1, 2, \ldots, 29$, and each positive integer x, we have $N_i(x) \geqslant \frac{x}{e}$ ($e = 2.71828\cdots$), prove that there exist i, j ($1 \leqslant i < j \leqslant 29$), such that $N_{ij}(1988) > 200$. (29th IMO alternative)

(2) Let A_i be subsets of $M = \{1, 2, \ldots, 10\}$, and $|A_i| = 5$ ($i = 1, 2, \ldots, k$), $|A_i \cap A_j| \leqslant 2$ ($1 \leqslant i < j \leqslant k$). Find the maximum value of k. (1994 China National Team Test)

(3) Assume that X is a finite set, $A_1, A_2, \ldots, A_m$ are subsets of X, and $|A_i| = r$ ($1 \leqslant i \leqslant m$). If for any $i \neq j$, we have $|A_i \cap A_j| \leqslant k$, prove that: $|X| \geqslant \dfrac{mr^2}{r + (m-1)k}$.

(4) There are 16 students in a test. Each question is a multiple choice question, and has four choices. After the test, it is found that for any two students there is at most one question for which their answers are the same. At most how many questions can there be? (33rd IMO China National Team Selection)

(5) There are k 5-element subsets $A_1, A_2, \ldots, A_k$ of set $M = \{1, 2, \ldots, 10\}$, that satisfy the condition: any two elements in M appear in at most two of the subsets A_i, A_j ($i \neq j$). Find the maximum value of k.

(6) There are n airports in a nation, and k airlines. There are nonstop flights between some airports (which are two-ways). If there is no nonstop flight between two airports, it is always able to fly to B from A by connecting flights.

(a) It is known that to connect these n airports, we need $n - 1$ routes (this can be used without proof). Since airlines often close down, they have to cooperate to make sure that the routes are always connected. Now, to make sure that when one airline closes down, the rest airlines can still provide flights between any two airports, how many nonstop flights should these airlines provide at least?

(b) When $n = 7$, $k = 5$, to make sure that when any two airlines close down, the rest airlines can still provide flight between any two airports, how many nonstop flights should these airlines provide at least?

In some extremization problems there are many variables, which make the problem seemingly complicated. However, by introducing new parameters, we can represent the target function using the parameters. Thus, the discrete extremization problem can be reduced to an equation with one or two unknown variables, which simplifies the problem. This approach to finding extrema is called parameter estimation.

Example 1. In one competition there are 20 gymnasts and 9 referees. Each referee ranks the gymnasts from 1 to 20 after their performance. It is known that the differences between the nine positions given by the referees for each gymnast are at most 3. Now for each gymnast, consider the sum of the nine ranks he/she gets, and rearrange these sums as $c_1 \leqslant c_2 \leqslant \cdots \leqslant c_{20}$. Find the maximum value of c_1. (The 2nd Soviet Mathematical Olympiad)

Analysis and Solution. First notice that we want to prove $c_1 \leqslant P$, which is equivalent to finding an i such that $c_i \leqslant P$. Thus, we can estimate c_1 by estimating the score (sum of ranks) of each player.

To minimize the score of a player, we should let more referees rank him or her as No. 1. Thus, we can estimate the score of the player who has the most number of No. 1 ranks. In one situation the estimate is obvious; that is when the player has nine No. 1 rankings. In this case we have $c_1 \leqslant 9$. Now, when the No. 1 rankings "*distribute widely*", we should use global estimation. Consider the sum of scores of players who have got at least one No. 1 ranking. Assume that there are r such players.

(1) $r = 1$, then $c_1 = 9$.

(2) $r = 2$. Since there are nine No. 1 rankings, at least one player A must have five No. 1 rankings. Since the differences between rankings from the same player are no more than 3, we know that the other four rankings of A should be no more than 4. Thus A has a score that is no more than $5 + 4 \times 4 = 21$. This means $c_1 \leqslant 21$.

(3) $r = 3$, consider the sum S of the scores of these players. There are 9 No. 1 rankings, and 18 other rankings which are no more than 4. Thus $S \leqslant 9 + 9 \times 3 + 9 \times 4 = 72$, so $c_1 \leqslant \dfrac{72}{3} = 24$.

(4) $r = 4$, similarly, we have $S \leqslant 90$, thus $c_1 \leqslant \left\lceil \dfrac{90}{4} \right\rceil = 22$.

(5) $r \geqslant 5$, then each ranking of these r people should be no more than 4, so there are at least $5r \geqslant 5 \times 9 = 45$ rankings which are no more than 4. However, the nine referees can only provide $9 \times 4 = 36$ rankings from 1 to 4, which is a contradiction.

Finally, $c_1 = 24$ is possible.

Let

$$c_1 = c_2 = c_3 = (1 + 1 + 1) + (3 + 3 + 3) + (4 + 4 + 4),$$
$$c_4 = (2 + 2 + 2 + 2 + 2) + (5 + 5 + 5 + 5),$$
$$c_5 = (2 + 2 + 2 + 2) + (5 + 5 + 5 + 5 + 5),$$

and $c_i = i + i + \cdots + i$ $(i = 6, 7, 8, \ldots, 20)$, as the table below shows.

judge \ player	1	2	3	4	5	6	7	8	9	...	20
1	1	3	4	2	5	6	7	8	9	...	20
2	1	3	4	2	5	6	7	8	9	...	20
3	1	3	4	2	5	6	7	8	9	...	20
4	3	4	1	2	5	6	7	8	9	...	20
5	3	4	1	2	5	6	7	8	9	...	20
6	3	4	1	5	2	6	7	8	9	...	20

Continued

judge \ player	1	2	3	4	5	6	7	8	9	...	20
7	4	1	3	5	2	6	7	8	9	...	20
8	4	1	3	5	2	6	7	8	9	...	20
9	4	1	3	5	2	6	7	8	9	...	20
sum of places	24	24	24	30	33	54	63	72	81	...	180

So the maximum value of c_1 is 24.

Example 2. There are r people participating in a chess tournament, where any two people play against each other once. The winner of each game gains two points, the loser gains zero point, and each gains one point in the case of a draw. After the tournament, there is only one player who has the least number of wins and the highest score. Find the minimum value of r. (16th Russian Mathematical Olympiad)

Analysis and Solution. First we consider the condition that only one player has the least number of wins and the highest score, and assume that this player is A. To calculate the score of A, we assume that A wins n games and draws m games, then A has $2n + m$ points.

To use the condition that A has the highest score, we should calculate the score of other players. Thus we should know the number of wins, losses and draws for other players, where the condition that A has the least number of wins can be used. Since any other player wins at least $n + 1$ games, and thus gaining at least $2n + 2$ points, we know that $2n + 2 < 2n + m$, $m > 2$, so $m \geqslant 3$.

This estimation is not optimal, but using it we can find a player with a higher score. Since $m \geqslant 3$, there are at least three players that tie with A. Assume one of them is B, then B has at least $(2n + 2) + 1 = 2n + 3$ points.

Thus $2n + 3 < 2n + m$, $m > 3$, so $m \geqslant 4$.

Next we prove that $n > 0$. If $n = 0$, then A wins no game. Thus the score of A is $S(A) \leqslant r - 1$. However, the total scores of all r people is $2C_r^2 = r(r-1)$, so the average score is $r - 1$. Since A has the highest score, we know $S(A) = r - 1$ and each one has score $r - 1$, which is a contradiction.

In conclusion, we have $r \geqslant (m+n) + 1 \geqslant 6$.

Finally, $r = 6$ is possible, as shown in the following table.

	A	B	C	D	E	F	total points
A		1	1	1	1	2	6
B	1		2	0	0	2	5
C	1	0		0	2	2	5
D	1	2	2		0	0	5
E	1	2	0	2		0	5
F	0	0	0	2	2		4

Thus the minimum value of r is 6.

Example 3. $n(n \geqslant 5)$ football teams participate in a single round-robin tournament. Each two teams play once, and the winner gets three points, the loser gets 0 point, and both get one point in case of a draw. After the tournament, the third last team has a score strictly lower than any team before it, and strictly higher than the two teams after it; it also has strictly more wins than the teams before it, and strictly less wins than the two teams after it. Find the minimum value of n.

Solution. Let $A_1, A_2, \ldots, A_{n-3}$ be the teams with higher score, B be the third last team, and C_1, C_2 be the two teams with lower score. Assume that B has x wins, y draws and z losses, then $n = x + y + z + 1$. Since C_1 has at least $(x+1)$ wins, we know $3x + y \geqslant 3(x+1) + 1$, so $y \geqslant 4$. Since A_1 has at most $(x-1)$ wins and at most $(n-x)$ draws, we have $3x + y + 1 \leqslant 3(x-1) + (n-x) = 3x + y + z - 2$, thus $z \geqslant$

3. That is, B has at least three losses. So there is at least one team in $A_1, \ldots, A_{n-3}$ which wins B, thus $x \geqslant 2$.

(1) If all A_i have drawn with C_1, C_2, B, then since $y = n - x - z - 1 \leqslant n - 6$, there are at least three teams in these $(n - 3)$ teams which draw with one of C_1, C_2. Thus one of C_1, C_2 has at least two draws, so $3x + y \geqslant 3(x + 1) + 2 + 1$, $y \geqslant 6$, so $n \geqslant 12$. Also the equality does not hold, since otherwise $z = 3$, A_i does not lose and has at most $x - 1 = 1$ wins, thus it has at least 10 draws. Therefore B, C_1, C_2 have at least 18 draws in total, but since $y = 6$, C_1 and C_2 can have at most two draws, so $18 \leqslant 6 + 2 + 2$, which is a contradiction. So $n \geqslant 13$.

(2) If some A_i (say A_1) has no draw with B, C_1, C_2, then it has at most $(n - 4)$ draws. Thus $3x + y + 1 \leqslant 3(x - 1) + (n - 4)$, $n \geqslant y + 8 \geqslant 12$. Equality again cannot hold, since otherwise $y = 4$, and from the previous arguments, we can see that C_1 and C_2 have no draw, thus A_i has at least $(x - 1)$ wins or it has at least 11 draws while C_1 has no draw, which is a contradiction. So A_i has $(x - 1)$ wins, eight or nine draws. Assume that k of the A_i's have eight draws and $(9 - k)$ of them have nine draws.

Note that C_1, C_2 have $(x + 1)$ wins and $(10 - x)$ losses, we have that $9(x - 1) + x + 2(x + 1) = k(4 - x) + (9 - k)(3 - x) + 7 - x + 2(10 - x)$, or $24x = k + 61$. For $0 \leqslant k \leqslant 9$, the equation has no integer solution, which is a contradiction. Thus $n \geqslant 13$.

From (1) and (2), we know $n \geqslant 13$ in any case.

On the other hand, when $n = 13$, we may construct the result of the matches as in the following table.

Thus $n_{\min} = 13$.

	A_1	A_2	A_3	A_4	A_5	A_6	A_7	A_8	A_9	A_{10}	B	C_1	C_2
A_1		3	1	1	1	1	1	1	1	1	1	0	3
A_2	0		1	1	1	1	1	1	1	1	1	3	3
A_3	1	1		3	1	1	1	1	1	1	1	0	3
A_4	1	1	0		1	1	1	1	1	1	1	3	3
A_5	1	1	1	1		1	1	1	1	1	3	0	3
A_6	1	1	1	1	1		1	1	1	1	3	3	0
A_7	1	1	1	1	1	1		1	1	1	3	3	0
A_8	1	1	1	1	1	1	1		1	1	3	3	0
A_9	1	1	1	1	1	1	1	1		1	3	3	0
A_{10}	1	1	1	1	1	1	1	1	1		0	3	3
B	1	1	1	1	0	0	0	0	0	3		3	3
C_1	3	0	3	0	3	0	0	0	0	0	0		3
C_2	0	0	0	0	0	3	3	3	3	0	0	0	

Example 4. There are 1000 certificates numbered from 000 to 999 and 100 boxes numbered from 00 to 99. If the number of a box is equal to the number of a certificate with one digit removed, then the certificate can be put in the box. If all the certificates can be put in k boxes, find the minimum value of k.

Analysis and Solution. First find a sufficient condition so that all of the certificates can be put in.

Consider all of the certificates containing digit a, b, c (a, b, c may be the same), to put them in boxes, the boxes used should contain 2 of the digits, say a and b. To make sure the certificates numbered with digits a, b, c in different order can be put in, a sufficient condition is that both boxes with numbers $\overline{ab}$ and $\overline{ba}$ are chosen. This suggests that, to guarantee that there are always two numbers chosen from any three numbers, and by any order, we may apply the pigeonhole principle: put three numbers in two sets, so there is at least one set with two numbers. Therefore, divide 0, 1, 2 , ..., 9 into two subsets A, B. If any two numbers in the same set are chosen with any order (all possible sequences with two repeatable digits), then this arrangement satisfies the requirements.

Let $| A | = k$, $| B | = 10 - k$. Then the numbers of 2-digit sequences with repeatable digits for A and B are k^2 and $(10 - k)^2$ respectively. So there are $k^2 + (10 - k)^2 = 2(k - 5)^2 + 50 \geqslant 50$ such sequences.

Let $k = 5$; that is, $| A | = | B | = 5$, e.g. $A = \{0, 1, 2, 3, 4\}$, $B = \{5, 6, 7, 8, 9\}$. Then there are 25 sequences each, so the resulting 50 boxes satisfy the requirements.

Next we prove that $k \geqslant 50$.

Assume that the number of chosen boxes beginning with 9 is the smallest, say m. Denote these boxes by $\overline{9a_i}(i = 1, 2, ..., m)$. Let $A = \{a_1, a_2, ..., a_m\}$, for any two digits a, b not in A, consider the certificate with number $\overline{9ab}$. Since a does not belong to A, then there is no box with number $\overline{9a}$. Similarly, there is no box with number $\overline{9b}$, so the box with number $\overline{ab}$ must be chosen. Note that a, b both have $10 - m$ possible values. Therefore, there are $(10 - m)^2$ such boxes, and that these boxes do not start with a_1, a_2, ..., a_m. By our choice of m, the numbers of boxes starting with a_1, a_2, ..., a_m should all be at least m. These determines are at least m^2 such boxes.

Thus, $k \geqslant m^2 + (10 - m)^2 \geqslant \dfrac{1}{2}[m + (10 - m)]^2 = 50$.

In conclusion, the minimum value of k is 50.

Example 5. Find the smallest natural number n with the following property: if one paints any five vertices of a regular n-polygon S red, there is always an axis of symmetry L of S, such that the symmetric point of each red point by L is not red.

Solution. Let the regular n-polygon be $A_1 A_2 \cdots A_n$, which has n axes of symmetry.

Denote the axis of symmetry across the midpoint of side $A_1 A_n$ by

L_1, the axis of symmetry across A_1 by L_2, the axis of symmetry across the midpoint of side $A_1 A_2$ by L_3, the axis of symmetry across A_2 by L_4, ..., the axis of symmetry across the midpoint of side $A_{\lceil \frac{n}{2} \rceil} A_{\lceil \frac{n+1}{2} \rceil}$ by L_n (where $A_{\lceil \frac{n}{2} \rceil}$ and $A_{\lceil \frac{n+1}{2} \rceil}$ can overlap). It is easy to know that, A_i and A_j are symmetric with respect to L_m, if and only if $i + j = m \pmod{n}$, where $A_{n+i} = A_i$.

For any set $X = \{a_1, a_2, \ldots, a_t\}$, define $X^* = \{a_i + a_j \mid 1 \leqslant i < j \leqslant t\}$. For $k \in \{3, 4, 5, \ldots, 13\}$, paint the five vertices A_1, A_2, A_4, A_6, A_7 red (where the subscripts are taken modulo k, and when there are less than five different vertices, repeatedly paint some random points red, so that there are five red points).

Let $P = \{1, 2, 4, 6, 7\}$, then $P^* = \{2, 3, 4, \ldots, 14\}$, so P^* contains a complete set of residuals modulo k (since it contains k continuous natural numbers). Thus, for any $m (1 \leqslant m \leqslant k)$, P^* contains an $x \equiv m \pmod{k}$, so there exists $i + j \equiv m \pmod{k}$, where i, $j \in P$. Thus A_i, A_j are two red points symmetric with respect to L_m. So, for any $k \in \{3, 4, 5, \ldots, 13\}$ the requirement is not satisfied. Therefore, $n \geqslant 14$.

On the other hand, for a regular 14-polygon, paint any of its five vertices A_{i_1}, A_{i_2}, ..., A_{i_5} red.

Let $P = \{i_1, i_2, \ldots, i_5\}$. Assume that there are r odd numbers and $5 - r$ even numbers in P, then the number of odd numbers in P^* is: $r(5-r) \leqslant \left[\left(\frac{5}{2} \right)^2 \right] = 6$ (the elements in P and P^* are taken modulo 14, which does not change the parity). Thus, there must exist an odd number m in $\{1, 3, 5, 7, 9, 11, 13\}$, such that $m \notin P^*$, thus for any $i, j \in P$, $i + j \not\equiv m \pmod{14}$. Thus any two red vertices are not symmetric with respect to L_m. So $n_{\min} = 14$.

Example 6. For an integer $n \geqslant 4$, find the minimum positive integer $f(n)$, such that for any positive integer m, in any subset of the set $\{m, m+1, \ldots, m+n-1\}$ with $f(n)$ elements, there exist at least three different elements which are relatively prime to each other. (2004

National Mathematical Olympiad in Senior)

Solution. When $n \geqslant 4$, for the set $M = \{m, m+1, m+2, \ldots, m+n-1\}$, if m is odd, then m, $m+1$, $m+2$ are relatively prime to each other; if m is even, then $m+1$, $m+2$, $m+3$ are relatively prime to each other. Thus there are at least three elements relatively prime to each other in any subset of M with n elements. So $f(n)$ exists, and $f(n) \leqslant n$.

Let $T_n = \{t \mid t \leqslant n+1\}$, and $2 \mid t$ or $3 \mid t$, then T_n is a subset of $\{2, 3, \ldots, n+1\}$, and any three numbers in T_n are not relatively prime with each other, so $f(n) \geqslant \mid T_n \mid + 1$.

By the inclusion-exclusion principle, $\mid T_n \mid = \left[\dfrac{n+1}{2}\right] + \left[\dfrac{n+1}{3}\right] - \left[\dfrac{n+1}{6}\right] + 1$, so

$$f(n) \geqslant \left[\frac{n+1}{2}\right] + \left[\frac{n+1}{3}\right] - \left[\frac{n+1}{6}\right] + 1.$$

Besides, notice that $\{m, m+1, m+2, \ldots, m+n\} = \{m, m+1, m+2, \ldots, m+n-1\} \cup \{m+n\}$, so $f(n+1) \leqslant f(n) + 1$.

So $f(4) \geqslant 4$, $f(5) \geqslant 5$, $f(6) \geqslant 5$, $f(7) \geqslant 6$, $f(8) \geqslant 7$, $f(9) \geqslant 8$.

Next we prove that $f(6) = 5$.

Let $x_1, x_2, x_3, x_4, x_5 \in \{m, m+1, m+2, \ldots, m+5\}$, and assume that k of the $x_i (1 \leqslant i \leqslant 5)$ are odd.

Since there are at most three even numbers in $\{m, m+1, m+2, \ldots, m+5\}$, we know that there are at most three even numbers in x_1, x_2, x_3, x_4, x_5, thus $k \geqslant 2$.

Since there are at most three odd numbers in $\{m, m+1, m+2, \ldots, m+5\}$, we know that there are at most three odd numbers in x_1, x_2, x_3, x_4, x_5, thus $k \leqslant 3$.

If $k = 3$, then the three odd numbers are relatively prime to each other.

If $k = 2$, assume that x_1, x_2 are odd, and x_3, x_4, x_5 are even. When $3 \leqslant i < j \leqslant 5$, $\mid x_i - x_j \mid \leqslant (m+5) - m = 5$. Since $x_i - x_j$ is even, $\mid x_i - x_j \mid = 2$ or 4. Thus it is not possible that $x_i \equiv x_j \pmod 3$ or $x_i \equiv x_j \pmod 5$, so there is at most one multiple of 3 in x_3, x_4, x_5, and at most one multiple of 5, thus there is at least one number which is neither a multiple of 3 nor a multiple of 5. Assume that it is x_3.

Consider the three numbers x_1, x_2, x_3. When $1 \leqslant i < j \leqslant 3$, $\mid x_i - x_j \mid \leqslant (m + 5) - m = 5$, so the greatest common factor of x_i and x_j is no more than 5. Since x_3 is neither a multiple of 3 nor a multiple of 5, we know that $(x_1, x_3) = (x_2, x_3) = 1$. Since $x_1 - x_2$ is even, we know that $\mid x_1 - x_2 \mid = 2$ or 4.

Thus $(x_1, x_2) = 1$, so x_1, x_2, x_3 are relatively prime to each other. Thus $f(6) = 5$.

Since $f(7) \geqslant 6$, $f(7) \leqslant f(6) + 1 = 5 + 1 = 6$, we have $f(7) = 6$.

Similarly, $f(8) = 7$, $f(9) = 8$.

This means that when $4 \leqslant n \leqslant 9$,

$$f(n) = \left[\frac{n+1}{2}\right] + \left[\frac{n+1}{3}\right] - \left[\frac{n+1}{6}\right] + 1. \qquad ①$$

Next we prove by induction that ① holds for all positive integers larger than 3.

Assume that ① holds when $n \leqslant k$ ($k \geqslant 9$). When $n = k + 1$, since

$$\{m, m+1, m+2, \ldots, m+k\} = \{m, m+1, m+2, \ldots, m +k - 6\} \cup \{m+k-5, m+k-4, \ldots, m+k\},$$

we have $f(k+1) \leqslant f(k-5) + f(6) - 1$. From the assumption, we have

$$f(k+1) \leqslant \left(\left[\frac{k-5+1}{2}\right] + \left[\frac{k-5+1}{3}\right] - \left[\frac{k-5+1}{6}\right] + 1\right)$$

$$+ \left(\left[\frac{6+1}{2}\right] + \left[\frac{6+1}{3}\right] - \left[\frac{6+1}{6}\right] + 1\right) - 1$$

$$= \left(\left[\frac{k}{2}\right] - 2 + \left[\frac{k-1}{3}\right] - 1 - \left[\frac{k-4}{6}\right] + 1\right) + (3 + 2 - 1 + 1) - 1$$

$$= \left(\left[\frac{k}{2}\right] + 1 + \left[\frac{k-1}{3}\right] + 1 - \left[\frac{k-4}{6}\right] - 1\right) + 1$$

$$= \left[\frac{k+2}{2}\right] + \left[\frac{k+2}{3}\right] - \left[\frac{k+2}{6}\right] + 1.$$

So for all positive integers n larger than 3, we have

$$f(n) = \left[\frac{n+1}{2}\right] + \left[\frac{n+1}{3}\right] - \left[\frac{n+1}{6}\right] + 1.$$

Example 7. There are n teams in a football league. Each team has two kinds of kits, one for home games and one for away games. In each match, if the colors of home kits for both teams are different, then both teams will wear home kits; otherwise the home team will wear the home kit, and the away team will wear the away kit. It is known that any two matches between four different teams involve at least three different kit colors in total. How many different kit colors must there be so that this can be true? (2010 China National Team Test)

Solution. At least $n - 1$ colors are needed.

First we construct an example with $n - 1$ colors used. Assume that the $n - 1$ colors are C_1, C_2, ..., C_{n-1}, and the n teams are T_1, T_2, ..., T_n. Let teams T_1, T_2 both use C_1 as the color of their home kits, C_2 as their away kits; the home kit of team $T_i (3 \leq i \leq n)$ uses color C_{i-1}, the away kit of team T_i uses color C_{i-2}.

By the rule of the problem, in each match, the color of home kits of both teams will appear. For any two games between four different teams, their home kits contain at least three different colors, which will appear in the two games. Thus this satisfies the requirements.

Next, assume that we can use at most $n - 2$ colors, so that at least three different colors appear in any two games between four different teams. Assume that exactly $n - 2$ colors have been used, and that these $n - 2$ colors are C_1, C_2, ..., C_{n-2}. Let x_i be the number of teams with color C_i, then $\sum_{i=1}^{n-2} x_i = n$. By the pigeonhole principle, there is at least one $x_i \geq 2$, say $x_1 \geq 2$.

If there is another $x_j \geq 2$, then assume that a, c are two teams with home kit color C_1, b, d are two teams with home uniform color C_j, then in the match between a (home) and b (away) and the match between c (home) and d (away), the four teams will all wear their home kits, and their kits all have color C_1 or C_j, which is a contradiction. Thus the rest x_j are all 0 or 1.

Thus $x_1 \geqslant n - ((n-2) - 1) = 3$.

If $x_1 = 3$, then $x_2 = x_3 = \cdots = x_{n-2} = 1$. Assume that the three teams with C_1 as the home kit color are a, b, c, and pick another team d, such that the home kit color of d is the same with the away kit color of b. (Since $x_2 = x_3 = \cdots = x_{n-2} = 1$, it is always possible.) Then, in the match between a (home) and b (away) and the match between c (home) and d (away), only team b wears the away kit, and there are only two kit colors, which is a contradiction.

If $x_1 \geqslant 4$, then consider the away kits of all the teams whose home kit color is C_1, and the home kits of other teams. There are $n - 2$ colors to choose for these n kit colors. By the pigeonhole principle, there are at least two types of kits with the same color. Since we have proved previously that each x_j other than x_1 must be 0 or 1, we must have one of the two scenarios below:

Scenario 1: there are two teams with home kit color C_1, and have the same away kit color. Assume that the two teams are b, d, find another two teams a, c with home kit color C_1. Then in the match between a (home) and b (away) and the match between c (home) and d (away), b, d wear away kits, and there are only two kit colors, which is a contradiction.

Scenario 2: there is a team whose home kit color is the same as the color of the away kit of another team. Assume that the former one is b and the latter one is d. Choose another two teams a, c which have the same home kit color C_1. Then, in the match between a (home) and b (away) and the match between c (home) and d (away), only b wears away kit. There are thus only two different colors, which is a contradiction.

Thus, there are at least $n - 1$ colors used in these kits.

Exercise 10

(1) There are 30 multiple choice questions in a test. One gains five points for each question answered correctly, zero point for each equation answered incorrectly, and one point for each equation not answered. Person A has gained more than 80 points, and tells B her score. From this information, B can infer the number of questions A has answered correctly. If A has a lower score which is still larger than 80, then B cannot infer the number of questions A has answered correctly. What is the score of A in this test? (The 2nd America Mathematics Invitational Tournament)

(2) There are 47 students in a class. The classroom has six rows, and each row has eight seats. Denote by (i, j) the seat in ith row and jth column. At the beginning of a new semester, the seats will be adjusted. Assume that the student who was in seat (i, j) before adjustment is in seat (m, n) after adjustment, we say that the student has a movement $[a, b] = [i - m, j - n]$, and call $a + b$ as the "*position number*" of such student. The sum of all the "*position numbers*" is denoted by S. Find the difference between the maximum and minimum values of S. (2003 Women Mathematical Olympiad)

(3) There are 12 troupes joining a show for a period of seven days. It is required that each troupe can see the show of the rest of the troupes, and that only the troupes without performance in that very day can watch the show. At least how many performances should be prepared in the show? (1994 China National Team Training)

(4) For each positive integer n, denote by $s(n)$ the largest integer satisfying the following requirements: for any natural number $k \leqslant s(n)$, n^2 can be represented as the square sum of k positive integers.

(i) Prove that: $s(n) \leqslant n^2 - 14 \ (n \geqslant 4)$.

(ii) Find an n such that $s(n) = n^2 - 14$.

(iii) Prove that: there are infinitely many n, such that $s(n) = n^2 - 14$. (33rd IMO)

(5) Given that real numbers $a_1, a_2, \ldots, a_n \ (n > 3)$ satisfy $a_1 +$

$a_2 + \cdots + a_n \geq n$, and $a_1^2 + a_2^2 + \cdots + a_n^2 \geq n^2$. Find the minimum value of $\max\{a_1, a_2, \ldots, a_n\}$. (28th America National Mathematical Olympiad)

Chapter 11   Counting in Two Ways

A typical method in combinatorial arguments is counting in two ways. The basic idea is as follows:

For a set $X = \{a_1, a_2, \ldots, a_n\}$, let $F = \{A_1, A_2, \ldots, A_k\}$ be a collection of subsets of X (usually a partition with nonempty intersections), and the ith subset A_i has r_i elements $(i = 1, 2, \ldots, k)$.

We now count some quantity Ω (called intermediate quantity) in two ways. On the one hand, we may count the contribution of each element to Ω, thus obtaining $\Omega = f(n)$. On the other hand, we may count the contribution of each subset A_i to Ω. Let this be $|\Omega_i| = f_i(k, r)$, then

$$\sum_{i=1}^{n} f_i(k, r) = \sum_{i=1}^{n} |\Omega_i| \leqslant |\Omega| = f(n).$$

Here we require that the contributions of the different subsets A_i, A_j are in some sense "*disjoint*", so we can add them up. Thus the intermediate quantity Ω should be carefully chosen to satisfy this requirement. If the different subsets A_i, A_j have contributions in common, then the common contribution should be removed.

The key to using the above-mentioned technique is to determine the quantity one counts. There is no general pattern in choosing the intermediate quantity, but usually we have the following ways to choose it.

Angle. When the problem is about monochromatic triangles, we can count the number of monochromatic angles or heterochromatic angles. It is an "*element reducing*" strategy.

r subsets. If the subsets satisfy the condition that $|A_i \cap A_j| \leqslant r$, then we may count on the $r + 1$-element subsets (the so-called *element increasing* strategy). In this case, for different subsets A_i, A_j, their $r + 1$ subsets must be different to each other, otherwise A_i and A_j would have a common $r + 1$-element subset, thus $| A_i \cap A_j | \geqslant r + 1$, which is a contradiction.

Pairs. We may pair two elements with special relations (instead of any two elements), and call it a pair. We then count the number of such pairs. Usually it is necessary to remove the repeated pairs.

Times. For example, the times of appearance of some element, and the person-time of attendance of an event.

Scores. The total score of a player, the sum of scores of all the players.

First we look at two examples in which we can apply the "*counting in two ways*" trick. Note that here counting is often a sub-problem of some extremization problems.

Example 1. There are 30 councilors in a council. Any two councilors are either friends or political opponents of each other. Each councilor has six political opponents. For a committee of three councilors, if any two of them are friends or any two of them are political opponents of each other, then the committee is called an "*odd*" committee. How many odd committees are there? (24th U.S.S.R. Mathematical Olympiad)

Solution. This question apparently has a graph theory flavor: political opponents or friends correspond to red edges or blue edges, each councilor having six political opponents corresponds to each point having six red lines and an "*odd*" committee corresponds to a monochromatic triangle.

Consider a complete graph of 30 vertices. Each point represents a councilor. If two councilors are political opponents, then draw a red line between the corresponding points, otherwise draw a blue line. From the assumption, each point has six red edges. Let us count the number ($| F |$) of monochromatic triangles (as subsets). We will apply the "*element reducing*" trick and choose the intermediate quantity to be the number S of monochromatic angles.

On the one hand, since starting from each point there are 6 red lines and 23 blue lines, we have $C_6^2 + C_{23}^2 = 268$ monochromatic angles. Thus we have $30 \times 268 = 8040$ monochromatic angles in total. The angles with different points as vertices are clearly different from each other. So $S = 8040$.

On the other hand, assume that there are x monochromatic triangles among all triangles. Then there are $C_{30}^3 - x = 4060 - x$ heterochromatic triangles. Each monochromatic triangle has three monochromatic angles, and each heterochromatic triangle has one monochromatic angle. Thus $S = 3x + 4060 - x = 2x + 4060$.

Since $3x + 4060 - x = 2x + 4060$, we have $x = 1990$.

Note. It is easier to count the number of heterochromatic angles. Since there are 0 heterochromatic angle in a heterochromatic triangle, we have

$$0 \times x + 2(4060 - x) = C_6^1 \times C_{23}^1 \times 30.$$

Example 2. There are 18 points in a plane, no three being collinear. Connect each pair of points with a segment, and color each segment red or blue. It is known that some point A has an odd number of red segments, and the other 17 points have different numbers of red segments.

(1) Find the number of red triangles in the graph.

(2) Find the number of triangles with exactly two red edges. (The 36th IMO alternative question)

Solution. Removing all of the blue lines from the graph, we get a simple graph G. From the condition, $d(A)$ is odd, and the degrees of the other points are different and belong to $\{0, 1, 2, \ldots, 17\}$. It is easy to understand that if a point has degree 17, then no point can have degree 0. So the degrees of other points are $0, 1, 2, \ldots, 16$ or $1, 2, \ldots, 17$.

In the former case, we have that $\sum d(x) = d(A) + 0 + 1 + 2 + \cdots +$ 16 is odd, which is impossible. So the degrees of the other points can only be 1, 2, ..., 17. Denote the other points by V_1, V_2, ..., V_{17}, and $d(V_i) = i (i = 1, 2, ..., 17)$.

Since $d(V_{17}) = 17$, V_{17} is connected to V_1. Since $d(V_1) = 1$, V_1 is only connected to V_{17}. Since $d(V_{16}) = 16$, V_{16} is connected to V_2, V_3, ..., V_{17}. Repeating this way, V_i is connected to V_{18-i}, V_{19-i}, ..., $V_{17} (i = 1, 2, ..., 8)$ and no others, and V_{18-i} is connected to every point except V_1, V_2, ..., $V_{i-1} (i = 1, 2, ..., 8)$. Now consider V_9. It should be connected to A, V_{17}, V_{16}, ..., V_{10}. Since V_9 is connected to A, we know that V_1, V_2, ..., V_8 are not connected to A, and V_9, V_{10}, ..., V_{17} are connected to A. Thus $d(A) = 9$.

Let $M = \{V_1, V_2, ..., V_8\}$, $N = \{A, V_9, V_{10}, ..., V_{17}\}$. Then the points in M are not connected to each other, and points in N are connected to each other, and point V_i in M is connected to exactly i points in N.

(1) The question is equivalent to finding the number of triangles in graph G. First, there are C_{10}^3 triangles in N. Next, for any point V_i in M, it is connected to i points in N, thus there are C_i^2 triangles. Therefore there are in total $\sum_{i=1}^{8} C_i^2 = C_9^3$ triangles, so the number of red triangles is $C_{10}^3 + C_9^3 = 204$.

(2) Count in two ways the total number of red angles.

On the one hand, since there are i edges starting from point V_i, the number of red angles with V_i as the vertex is C_i^2. So $S = C_9^2$ (red angles with A as the vertex) $+ \sum_{i=2}^{17} C_i^2 = C_9^2 + C_{18}^3$.

On the other hand, assume there are x triangles with exactly two red edges. Each of these triangles have exactly one red angle, thus we have x red angles. From (1), there are 204 red triangles, each of which having three red angles, thus we get $204 \times 3 = 612$ red angles. Apart from these, there are no red angles. Thus $S = x + 612$.

Since $C_9^2 + C_{18}^3 = S = x + 612$, we have $x = 240$.

Example 3. Assume that $S = \{1, 2, \ldots, 15\}$. Choose n subsets A_1, $A_2, \ldots, A_n$ which satisfy the following conditions:

(1) $|A_i| = 7$ $(i = 1, 2, \ldots, n)$,

(2) $|A_i \cap A_j| \leqslant 3$ $(1 \leqslant i < j \leqslant n)$,

(3) For any subset M of S with three elements, there is an A_k such that $M \subset A_k$.

Find the minimum number n of such subsets. (China National Team Selection Examination of 40th IMO)

Analysis and Solution. The condition (1) that $|A_i| = 7 (i = 1, 2, \ldots,$ $n)$ suggests us to count the total number S_1 of the appearance of elements in each subset.

On the one hand, counting within each subset, we have $S_1 = 7n$. On the other hand, counting with each element, assuming that i appears r_i times $(i = 1, 2, \ldots, 15)$, we have $S_1 = \sum_{i=1}^{15} r_i$. Thus $7n = S_1 = \sum_{i=1}^{15} r_i$. Next we find the range of $\sum_{i=1}^{15} r_i$, which would follow if we can find the range of each r_i. Without loss of generality, consider the range of r_1.

Now condition (3) suggests us to evaluate the number S_2 of subsets which contain 1 and have three elements.

On the one hand, there are $C_{14}^2 = 91$ ways to choose such a subset, so $S_2 = C_{14}^2 = 91$; on the other hand, consider the r_1 subsets containing 1, each contributing $C_6^2 = 15$ 3-element subsets containing 1, resulting in $15r_1$ subsets. By condition (3), these cover all the 91 3-element subsets containing 1, so we have $15r_1 \geqslant S_2 = 91$, $r_1 \geqslant 7$; similarly for $i = 1, 2, \ldots, 15$, we have $r_i \geqslant 7$, so $7n = \sum_{i=1}^{15} r_i \geqslant \sum_{i=1}^{15} 7 = 15 \times 7$, $n \geqslant 15$.

When $n = 15$, we can choose $A_i = \{1 + i - 1, 2 + i - 1, 4 + i - 1,$ $5 + i - 1, 6 + i - 1, 11 + i - 1, 13 + i - 1\}(i = 1, 2, \ldots, 15)$, where everything is taken modulo 15. We can verify that this arrangement satisfies all the requirements. Thus the minimum of n is 15.

Example 4. Eight singers are attending a concert. Suppose we want to arrange for m shows such that four of the eight singers perform in each show, and that each two of the eight singers appear together in some same number of shows. Find an arrangement for which m is minimal. (1996 China Mathematical Olympiad)

Analysis and Solution. First note that there are m shows, and four of the eight singers perform in each show; this is equivalent to having m 4-element subsets of a set with eight elements. Moreover, the condition that any two singers appear together some same number of times (say r) is equivalent to the condition that each 2-element subset appears r times in these 4-element subsets. This leads us to consider the total number S of appearances of the singers.

On the one hand, there are m shows and each involves four singers, thus $S = 4m$.

On the other hand, suppose that each two singers appear together precisely r times, thus we have $r\mathrm{C}_8^2$ pairs, which contributes $2r\mathrm{C}_8^2$ appearances to S. However, each appearance of some singer A is counted three times (since there are other three singers that appear together with A), thus $S = \dfrac{2r\mathrm{C}_8^2}{3}$, thus $4m = S = \dfrac{2r\mathrm{C}_8^2}{3}$, so $6m = r\mathrm{C}_8^2$, $3m = 14r$. Therefore $3 \mid r$, $r \geqslant 3$. Thus $3m = 14r \geqslant 14 \times 3 = 42$, so $m \geqslant 14$.

When $m = 14$, we can have the following arrangement. Thus the minimum value of m is 14.

A	1234	1256	1278	1357	1368	1458	1467
A'	5678	3478	3456	2468	2457	2367	2358

Note: If we compute the total number T as pairs, we can avoid repeated counting.

On the one hand, there are m shows and each involves four singers (and thus $\mathrm{C}_4^2 = 6$ pairs), so $T = m\mathrm{C}_4^2 = 6m$; on the other hand, each two singers appear together exactly r times, thus $= r\mathrm{C}_8^2$ also. We then argue as in the proof above.

Example 5. There are 10 birds on the ground, such that among any five birds, there are at least four birds that lie on some circle. Now, at least how many birds must lie on some same circle? (1991 China Mathematical Olympiad)

Analysis and Solution. Represent birds by points. Suppose that the largest number of points that lie on some same circle is r, then clearly $r \geqslant 4$. Can we have $r = 4$? We first discuss this.

If $r = 4$, then each circle contains at most four points. Since we are only considering that contains at least four points, we know that each such circle contains exactly four points.

Step 1: Compute the number of such "4-point circles". Note that any 5-point subset corresponds to a 4-point circle; we can try to do this using bijection. In fact, since any 5-point subset corresponds to a 4-point circle, we have $C_{10}^5 = 252$ 4-point circles in total. However, each 4-point circle belongs to precisely six 5-point subsets, and is thus counted six times. Therefore we have $\frac{252}{6} = 42$ 4-point circles. They must be different, since otherwise there are two different 4-point subsets $\{A, B, C, D\}$ and $\{A', B', C', D'\}$ that lie on the same circle. Noting that these two 4-point subsets involve at least five points in total, we obtain a circle containing at least five points, which is a contradiction.

Step 2: Study the partition and choose some appropriate intermediate quantity. The 42 4-point circles above can be seen as 42 different subsets, say $M_1, M_2, \ldots, M_{42}$. They satisfy that: $|M_i| = 4$, $|M_i \cap M_j| \leqslant 2$. Using the idea of "*element increasing*", we may consider the number S of three-element subsets.

On the one hand, $S = C_{10}^3 = 120$; on the other hand, we have $S \geqslant 42C_4^3 = 168$, which is a contradiction. Thus $r > 4$.

Let M be the circle containing the largest number of points. Since $r > 4$, there are at least five points A, B, C, D, E on M. Next we prove that at most one point is not on M.

In fact, suppose that P and Q are not on M, consider the five points P, Q, A, B, C. There are four concyclic points among them, but since P and Q are not on circle ABC, we may assume that P, Q, A, B lie on some circle M_1.

In the same way, considering P, Q, C, D, E, we may assume that P, Q, C, D lie on some circle M_2. Considering P, Q, A, C, E, we may assume that P, Q lie on the same circle with two points among A, C and E.

(i) If P, Q, A, C lie on some circle M_3, then M_3 coincides with M_1, thus P, Q, A, B, C are concyclic, which is a contradiction.

(ii) If P, Q, A, E lie on some circle M_3, then M_3 coincides with M_1, thus P, Q, A, B, E are concyclic, which is a contradiction.

(iii) If P, Q, C, E lie on some circle M_3, then M_3 coincides with M_2, thus P, Q, C, D, E are concyclic, which is a contradiction.

Therefore, we have $r \geqslant 9$. Also $r = 9$ is possible, since we can set that nine points lie on some circle, and the other point lies outside the circle. Therefore the minimum value of r is 9.

Note. This is a hard problem, but the average score of this problem in 1991 CMO was unexpectedly high. The reason is that the answer is easy to guess; once one proves $r > 4$, one can immediately find that $r = 9$, and the construction is also very easy.

One may also consider: what if the number 10 in the problem is replaced with n ?

Example 6. There are 16 students participating in an exam, where all problems are multiple choice problems with four choices. After the exam, it is found that for any two students there is at most one problem to which their answers are identical. At most how many problems can there be? (1992 China National Team Selection)

Analysis and Solution. Suppose that there are n problems. We prove that $n_{max} = 5$.

For each problem, the 16 answers to it for a sequence of 1, 2, 3, 4 with 16 terms, so that we obtain a $n \times 16$ table from these n

problems. Note that for any two students, there is at most one problem to which their answers are identical, we can compute the number of pairs formed by two identical answers to one problem. For simplicity, if two answers in the same row are identical, we will connect these two numbers by a segment. For each row, suppose that there are x $1's$, y $2's$, z $3's$ and w $4's$, then there will be $C_x^2 + C_y^2 + C_z^2 + C_t^2 \geqslant C_4^2 + C_4^2 + C_4^2 + C_4^2 = 24$ in that segment. Thus there are at least $24n$ segments in total. However, the assumption in the problem implies that the horizontal projection of any two segments cannot be the same, thus we know that $24n \leqslant C_{16}^2 = 120$, $n \leqslant 5$.

If $n = 5$, then equality must hold. Thus we have:

In each row, there are 4 1's, 4 2's, 4 3's and 4 4's;

The horizontal projections of different segments do not coincide.

We may assume that the first row is $(1111 \quad 2222 \quad 3333 \quad 4444)$ and view them as four groups, then in any other row, each group must be a permutation of 1, 2, 3, 4. For simplicity, we may set as many groups to the natural permutation (1234) as possible. After attempting, we find that the groups in the second row and the first four columns can all be set to (1234). Continuing in this way, we get a table that satisfies the requirements, showing that $n = 5$ is possible.

Problem No. \ Student	1	2	3	4	5	6	7	8	9	10	11	12	13	14	15	16
1	1	1	1	1	2	2	2	2	3	3	3	3	4	4	4	4
2	1	2	3	4	1	2	3	4	1	2	3	4	1	2	3	4
3	1	2	3	4	4	3	2	1	3	4	1	2	2	1	4	3
4	1	2	3	4	2	1	4	3	4	3	2	1	3	4	1	2
5	1	2	3	4	3	4	1	2	2	1	4	3	4	3	2	1

Note. The original solution used global estimates (see Exercise 9). Here we use instead the technique of counting in two ways, which is more natural and also helps to discover the correct construction.

Example 7. There are n people greeting each other by phone on a holiday. It is known that each one called at most three others, that each two people talked on the phone at most once, and that for any three people, at least one of them called one of the others. Find the maximum value of n. (Original)

Solution. We need the following lemma.

Lemma. If a simple graph G of order n does not contain K_3, then $\|G\| \leqslant \left[\frac{n^2}{4}\right]$.

Proof. Let A be the vertex with maximal degree.

Let $M = \{A_1, A_2, \ldots, A_r\}$ and $N = \{B_1, B_2, \ldots, B_s\}$ $(r + s + 1 = n)$ denote the set of points that are adjacent and not adjacent to A respectively. Since there is no triangle in G, we know that there cannot be any edge in M, thus all the edges of G are either in N or between M and N. Thus each edge has one vertex being B_1, B_2, $\ldots$, or B_s (see Figure 11.1).

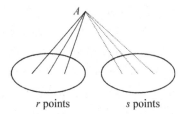

r points   *s* points

Figure 11. 1

Therefore,

$$\|G\| \leqslant d(A) + d(B_1) + d(B_2) + \cdots + d(B_s) \leqslant r + r + \cdots + r$$
$$= (s + 1)r \leqslant \left(\frac{s + r + 1}{2}\right)^2 = \frac{n^2}{4}.$$

Since $\|G\| \in \mathbf{Z}$, we know $\|G\| \leqslant \left[\frac{n^2}{4}\right]$. $\square$

Now we return to the original problem. Represent the n people by n points; if one person called another person, we will construct a directed edge between these two people. In this we form a directed graph G.

On the on hand, there is no triangle in $\bar{G}$, the complement of G; we know from the lemma that $\|\bar{G}\| \leqslant \left[\frac{n^2}{4}\right]$, thus $\|G\| = C_n^2 -$

$\|\bar{G}\| \geqslant C_n^2 - \left[\frac{n^2}{4}\right] = \left[\frac{(n-1)^2}{4}\right]$. On the other hand,

$$\| G \| = \sum_{i=1}^{n} d^{+}(x_i) \leqslant \sum_{i=1}^{n} 3 = 3n.$$

Therefore

$$\left[\frac{(n-1)^2}{4} \right] \leqslant 3n. \qquad \text{①}$$

When n is odd, ① is equivalent to $\frac{(n-1)^2}{4} \leqslant 3n$, thus $n \leqslant 13$;

when n is even, ① is equivalent to $\frac{n^2 - 2n}{4} \leqslant 3n$, thus $n \leqslant 14$.

In any case we have $n \leqslant 14$.

Finally, $n = 14$ is possible. Construct two K_7's; for each of them, denote it by $A_1 A_2 \ldots A_7$, then we set the direction of the edges so that A_i points to A_{i+1}, A_{i+2}, A_{i+3} $(i = 1, 2, \ldots, 7, A_{i+7} = A_i)$, then this directed graph satisfies the requirements.

In fact, each point has exactly three outgoing edges; also for any three points, we must have two points belonging to a same K_7. Let them be A_i, A_j, we can see that either A_i called A_j, or A_j called A_i, thus the requirements are met.

Therefore the maximum of n is 14.

Example 8. Three people, A, B, and C are playing table tennis. When two of them play against each other, the third person will be the referee; after one game, the winner goes on to play against the referee, and the loser becomes the referee for the next game. At some point we know that A has played a games, and B has played b games. Find the minimum number of games C must have played. (Original)

Solution. Suppose C has played c games, then we have $a + b + c$ games in total. Since each game involves two people, we know that the total number of games is $\frac{a + b + c}{2}$.

Therefore, the number of games in which C is the referee is

$$\frac{a + b + c}{2} - c = \frac{a + b - c}{2}.$$

Note that if C is the referee for some game, she must be one of the players for the next game; thus she cannot be the referee for any two consecutive games. Therefore $\frac{a+b-c}{2} \leqslant c+1$ (she can only be the referee during the $c+1$ "*gaps*" between the c games she played), so $c \geqslant \frac{a+b-2}{3}$.

Since c is an integer, we have $c \geqslant \left\lceil \frac{a+b-2}{3} + \frac{2}{3} \right\rceil = \left\lceil \frac{a+b}{3} \right\rceil$.

When $a+b=3k$, $c \geqslant \left\lceil \frac{a+b}{3} \right\rceil = k$. We may choose $a = 2k$, $b = k$, and denote any game in which A plays with B by (A, B, C); then one possibility where C plays exactly k games is (A, B, C), (A, C, B), (A, B, C), (A, C, B), ..., (A, B, C), (A, C, B), with $2k$ games in total.

When $a+b = 3k+1$, we have $c \geqslant \left\lceil \frac{a+b}{3} \right\rceil = k$. Since $c \geqslant \left\lceil \frac{a+b}{3} \right\rceil = k$, we have $c \neq k$, thus $c \geqslant k+1 = \left\lceil \frac{a+b}{3} \right\rceil + 1$. We may choose $a = 2k$, $b = k+1$, and one possibility is (A, B, C), (A, C, B), (A, B, C), (A, C, B), ..., (A, B, C), (A, C, B), $\underline{(B, C, A)}$, with $2k+1$ games in total and C has played $(k+1)$ games.

When $a+b = 3k+2$, $c \geqslant \left\lceil \frac{a+b}{3} \right\rceil = k$. Let $a = 2k+1$, $b = k+1$, one possibility is (A, B, C), (A, C, B), (A, B, C), (A, C, B), ..., (A, B, C), (A, C, B), (A, B, C), where there are $2k+1$ games in total, and $c = k = \left\lceil \frac{a+b}{3} \right\rceil$.

Therefore

$$c \min = \begin{cases} \left\lceil \dfrac{a+b}{3} \right\rceil & (a+b \equiv 0, 2 \pmod 3), \\ \left\lceil \dfrac{a+b}{3} \right\rceil + 1 & (a+b \equiv 1 \pmod 3). \end{cases}$$

Example 9. Suppose that X is a set with 56 elements. It is known that, for any 15 subsets of X, if the union of any 7 of them contains at least n elements, then we can always find three subsets from these 15 subsets such that their intersection is nonempty. Find the minimal value of n. (2006 China Mathematical Olympiad)

Solution. $n_{\min} = 41$.

First prove that $n = 41$ satisfies the requirements. Assume the contrary, that there exist 15 subsets of X such that the union of any 7 of them contains at least 41 elements, and that the intersection of any 3 of them is empty. We know that any element belongs to at most two subsets; by adding some elements to the sets, we may assume that each element belongs to exactly two subsets. By the pigeonhole principle, there must exist a set, say A, that contains at least $\left\lceil \dfrac{56 \times 2}{15} \right\rceil + 1 = 8$ elements. Also denote the other sets by $A_1, A_2, \ldots, A_{14}$.

Consider any seven subsets not including A, by assumption we know that their union contains at least 41 elements. Therefore we obtain $41 C_{14}^7$ elements in total.

On the other hand, for any element a, if $a \notin A$, then precisely two of the sets $A_1, A_2, \ldots, A_{14}$ contain a, thus a is counted $(C_{14}^7 - C_{12}^7)$ times above; if $a \in A$, then precisely one of the sets $A_1, A_2, \ldots,$ A_{14} contains a, thus a is counted $(C_{14}^7 - C_{13}^7)$ times above.

Therefore $41 C_{14}^7 \leqslant (56 - |A|)(C_{14}^7 - C_{12}^7) + |A| \cdot (C_{14}^7 - C_{13}^7) = 56(C_{14}^7 - C_{12}^7) - |A|(C_{13}^7 - C_{12}^7) \leqslant 56(C_{14}^7 - C_{12}^7) - 8(C_{13}^7 - C_{12}^7)$, thus $48 C_{12}^7 + 8 C_{13}^7 \leqslant 15 C_{14}^7$, which simplifies to $3 \times 48 + 4 \times 13 \leqslant 15 \times 13$, or $196 \leqslant 195$, which is a contradiction.

Next we prove that $n \geqslant 41$, again by contradiction.

If $n \leqslant 40$, we may assume $X = \{1, 2, \ldots, 56\}$ and let $A_i = \{x \in X \mid x \equiv i \pmod{7}\}$ $(i = 1, 2, \ldots, 7)$, $B_j = \{x \in X \mid x \equiv j \pmod{8}\}$ $(j = 1, 2, \ldots, 8)$. Obviously $|A_i| = 8$, $|A_i \cap A_j| = 0 (1 \leqslant i < j \leqslant 7)$ and $|B_j| = 7$, $|B_i \cap B_j| = 0 (1 \leqslant i < j \leqslant 8)$. By Chinese Remainder Theorem, we know that $|A_i \cap B_j| = 1 (1 \leqslant i \leqslant 7, 1 \leqslant j \leqslant 8)$. Therefore, for any 3 of the 15 sets, at least two of them must both be

A_i or both be B_j, thus their intersection is empty. Moreover, for any 7 of the 15 subsets, suppose that t $(0 \leqslant t \leqslant 7)$ of them are A_i and $(7-t)$ of them are B_j, then by inclusion-exclusion principle we have that the cardinality of their union is

$$8t + 7(7 - t) - t(7 - t) = 49 - t(6 - t) \geqslant 49 - 9 = 40.$$

Therefore, for these 51 sets, the union of any 7 of them has at least 40 (thus n) elements, and the intersection of any 3 of them is empty, which is a contradiction.

Therefore, the minimum value of n is 41.

Exercise 11

(1) 1650 students are arranged in 22 rows and 75 columns. It is known that for any two columns, there are at most 11 pairs of students such that they are in these two columns respectively, that they are in the same row, and that they have the same gender. Prove that the number of males among these people does not exceed 928. (CWMO 2003)

(2) There are $12k$ people attending a meeting. It is known that each one has talked to precisely $(3k + 6)$ people among them, and that for any two people, the number of people that have talked to both of them is the same. Find all possible values of the number of people. (The 36th IMO alternative question)

(3) In a competition that lasts k days, there are n participants. It is known that in each day, the scores of the participants form a permutation of $1, 2, \ldots, n$. If at the end of day k, the total score of each participant is 26, find all possibility values of (n, k).

(4) There are 30 students in a class with different ages. Some of the students in the class are friends; the number of friends of each student is the same. We call a student A "*senior*" if A is older than more than half (not inclusive) of his or her friends. At most how

many "*senior*" students can there be? (The 20th Russian Mathematical Olympiad)

(5) Some students are taking an exam. There are four multiple choice questions, each having three choices. It is known that for any three students, there exists one problem to which their three answers are mutually different. Find the maximal number of students. (The 29th IMO alternative question)

(6) Among 25 people, any 5 of them may form a committee. It is known that any two committees have at most one common member; prove that there can be at most 30 committees. (The 5th Russian Mathematical Olympiad)

Chapter 12 Shrinking the Encirclement

In some problems we may use the following strategy. First we find a large *"encirclement"* which is a rough estimate for the target function, and then we narrow the *"encirclement"* by exploiting the requirements in the problem. In this process we may consider the various factors in the requirements and analyze their effects on estimating the target function. In this way we can try to improve the estimates (i.e. *"shrinking the encirclement"*), and finally obtain the optimal estimate.

Example 1. In an $n \times n$ chessboard C, each cell is filled with a number, in some particular order. First the cells on the boundary of the chessboard are filled with -1. Then the rest of the cells are filled in some order, such that each cell is filled with the number which is the product of the two numbers which are in either the same row or the same column of the cell, and closest to the cell. Find the maximum number $f(n)$ of 1s and the minimum number $g(n)$ of 1s. (The 20th USSR Mathematical Olympiad)

-1	-1	-1	-1	-1	-1
-1	a			b	-1
-1					-1
-1	e				-1
-1	d	f		c	-1
-1	-1	-1	-1	-1	-1

Analysis and Solution. First, apparently $f(3) = g(3) = 1$.

When $n > 3$, we hope that the number of 1s is as large as possible. Can all the cells be filled with 1? By trying, we find that there is at least one -1. Actually, consider the cells that are adjacent to some border cells. At least one of these cells is filled with -1; otherwise, consider the cells a, b, c, d as in the table. Assume that a, b, c are filled before d. Since a, b, c are filled with 1, d can only be filled with -1, which is a contradiction.

Thus, $f(n) \leqslant (n-2)^2 - 1$.

Next, by row product, a can be filled with 1. By column product, b is filled with 1. By row product again, c, e are filled with 1; by column product again, f is filled with 1, thus d is filled with -1. For the rest of the cells, the cells in the second row and the second last row can be filled with 1 by row product. The rest cells can be filled with 1 by column product. Thus there are $(n-2)^2 - 1$ cells filled with 1, so $f(n) = (n-2)^2 - 1$.

Next we prove that, when $n > 3$, $g(n) = n - 2$.

If $g(n) = r \leqslant n - 3$, then the r "1"s can occupy at most $n - 3$ rows and columns. Thus, in the $n - 2$ rows and $n - 2$ columns in the center, there must be one row and one column which contain only -1, which is impossible because the cell at the intersection of the row and column cannot be filled with -1. Thus $g(n) \geqslant n - 2$.

On the other hand, we may fill the $n - 2$ cells in the 2nd column (from above to below) with 1, and fill the cells with -1 in each row from left to right, so that there are $n - 2$ "1"s. Thus $g(n) = n - 2$.

Example 2. There are h 8×8 chessboards, and each board is filled with $1, 2, 3, \ldots, 64$ in each cell respectively, such that when any two of the boards overlap in any way, the numbers at the same position are different. Find the maximum value of h. (The 29th IMO alternative question)

Analysis and Solution. First look at the special cases. Consider a 2×2 chessboard. We find that any two cells in any two of the chessboards respectively may coincide. Thus the four cells in the same chessboard should be considered in the same class. Cells in the same class can only be filled with different integers.

So,

$$h \leqslant \left\lfloor \frac{\text{No. of integers}}{\text{No. of cells in the same class}} \right\rfloor = \left\lceil \frac{2 \times 2}{4} \right\rceil = 1,$$

so the maximum value of h is 1.

Consider the 3×3 chessboard. As the table on the right shows, the cells in the chessboard can be divided into three classes A, B, C, and any two cells in the same class may overlap. Thus the cells in the same class can only be filled with different integers. Note that there are four cells in the largest class. So $h \leqslant \left\lceil \frac{9}{4} \right\rceil = 2$.

$$
\begin{array}{ccc}
B & C & B \\
C & A & C \\
B & C & B
\end{array}
$$

When $h = 2$, the two chessboards can be filled as follows:

$$
\begin{array}{ccc}
2 & 6 & 3 \\
9 & 1 & 7 \\
5 & 8 & 4
\end{array}
\qquad
\begin{array}{ccc}
1 & 2 & 6 \\
5 & 9 & 3 \\
8 & 4 & 7
\end{array}
$$

So the maximum value of h is 2.

Consider then the 4×4 chessboard; the cells can be divided into four classes, where cells in the same class may overlap in different chessboards. So the numbers filled in the cells in the same class should be different. In this case, the largest class has four cells, so $h \leqslant \left\lceil \frac{16}{4} \right\rceil = 4$.

$$
\begin{array}{cccc}
B & C & D & B \\
D & A & A & C \\
C & A & A & D \\
B & D & C & B
\end{array}
$$

When $h = 4$, we only need to make sure that the cells in the same class are filled with different numbers. First fill the A cells in the four chessboards. The A cells in the first chessboard are filled with 1, 2, 3, 4 respectively. The A cells in the second chessboard are filled with 5, 6, 7, 8 respectively. The A cells in the third chessboard are filled with 9, 10, 11, 12 respectively. The A cells in the fourth chessboard are filled with 13, 14, 15, 16 respectively. Then fill the B cells of the four chessboards. The B cells in the first chessboard are filled with 5, 6, 7, 8 respectively. The B cells in the second chessboard are filled with 9, 10, 11, 12 respectively. The B cells in the third chessboard are filled with 13, 14, 15, 16 respectively. The B cells in the fourth chessboard are filled with 1, 2, 3, 4 respectively. The C, D cells are filled similarly. In other words, in the first chessboard, the A cells are filled with 1, 2, 3, 4; the B cells are filled with 5, 6, 7, 8; the C cells are filled with 9, 10, 11, 12; the D cells are filled with 13, 14, 15, 16. In the second chessboard, the numbers filled increase by 4, with respect to the numbers at the same position in the first chessboard (numbers larger than 16 are taken modulo 16), as the following table shows:

5	9	14	6		9	13	2	10		13	1	6	14		1	5	10	2
13	1	2	10		1	5	6	14		5	9	10	2		9	13	14	6
12	3	4	15		16	7	8	3		4	11	12	7		8	15	16	11
8	16	11	7		12	4	15	11		16	8	3	15		4	12	7	3

From the above discussion, we can see that the 8×8 chessboard can be divided into 16 different classes, represented by 16 letters A, B, ..., P (shown in the following figure). Cells in different classes will not coincide no matter how two chessboards overlap; cells in the same class may coincide. Thus cells in the same class should be filled with different numbers.

Each chessboard has four A cells, so $h \leqslant \left\lceil \dfrac{64}{4} \right\rceil = 16$. When $h = 16$, we will fill the chessboards so that numbers in the cells of the same class are different with each other. First we fill the 64 cells of the first chessboard. The A cells are filled with 1, 2, 3, 4; the B cells are filled with 5, 6, 7, 8, ... the P cells are filled with 61, 62, 63, 64. The numbers filled in the second chessboard are just numbers of the corresponding cells in the first board plus 4 (numbers larger than 64 are taken modulo 64). The numbers filled in the third chessboard are just numbers of the corresponding cells in the first board plus 8 (numbers larger than 64 are taken modulo 64), and so on. Thus $h_{max} = \left\lceil \frac{8 \times 8}{4} \right\rceil = 16$.

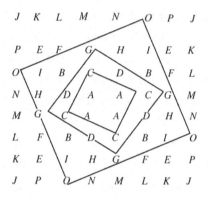

Another construction is as follows: divide the 64 cells into 16 groups: A_1, A_2, ..., A_{16}. Each group has four cells. For the first chessboard, the four cells in the first group are filled with 1, 2, 3, 4; the four cells in the second group are filled with 5, 6, 7, 8, ... the four cells in the 16th group are filled with 61, 62, 63, 64. For the ith chessboard, the numbers filled in the jth group in the $(i-1)$th chessboard are used to fill the $(j+1)$th group.

In general, for h chessboards, there are $4h$ cells which may overlap, thus they should be filled with different numbers. Thus $4h \leqslant n^2$, so $h \leqslant \frac{n^2}{4}$. Since $h \in \mathbf{Z}$, we have $h \leqslant \left\lceil \frac{n^2}{4} \right\rceil$. By similar construction as above, we can make $h = \left\lceil \frac{n^2}{4} \right\rceil$. So $h(n) = \left\lceil \frac{n^2}{4} \right\rceil$.

Example 3. There are 50 blanks in a lottery ticket. Each participant writes the numbers from 1 to 50 in the blanks (each number appears only once in the same ticket) of a lottery ticket. The tickets are then compared with the official ticket (the winning ticket). If a ticket has at least one position which is the same with the official ticket, then this ticket win the lottery. How many lotteries should a participant write, such that one of them is guaranteed to win? (The 25th USSR Mathematical Olympiad)

Analysis and Solution. We call an integer k a "*winning number*" if the participant can win the lottery by writing k lotteries properly. The k lotteries are arranged to a $k \times 50$ table:

$$a_1, a_2, a_3, \ldots, a_{50}$$
$$b_1, b_2, b_3, \ldots, b_{50}$$
$$\ldots$$
$$c_1, c_2, c_3, \ldots, c_{50}$$

Winning ticket $x_1, x_2, x_3, \ldots, x_{50}$.

That k is a winning number means that no matter how we write in the blanks x_1, x_2, x_3, $\ldots$, x_{50}, there is at least one column in the table which contains at least one number equaling x in this column. We call such number a "*good number*". Thus, k is a winning number means that there is at least one good number in the table.

Apparently, $k = 50$ is a winning number. Actually, the 50×50 table has 50 rows and 50 columns, where each row is a permutation of 1, 2, 3, $\ldots$, 50. There are 50 "1"s in the table. We can write the 50 "1"s in 50 different columns, thus each column has just one "1". Therefore, no matter what x_1, x_2, x_3, $\ldots$, x_{50} are, the "1" in these numbers must be in the same column with some "1" in the table. Furthermore, $k = 49$ is a winning number. We can construct the table as follows:

1	2	3	4	...	49	50
49	1	2	3	...	48	50
48	49	1	2	...	47	50
...	...	...	...	...	...	...
2	3	4	5	...	1	50

Since the last blank of the winning ticket can only be filled with one number, there must be a number i from $\{1, 2, \ldots, 49\}$ that is written in the previous blanks, and there must be a column which contains a number equaling to this number i. From this we find that, for any number i, we can include it in each column, so that in any column of the winning ticket, no matter what number is written, this number will equal some number in the same column of the table. This leads us to construct the following table:

$$1, 2, 3, \quad 4, \ldots, a-1, a$$
$$2, 3, 4, \quad 5, \ldots, \quad a, 1$$
$$3, 4, 5, \quad 6, \ldots, \quad 1, 2$$
$$\cdots$$
$$a, 1, 2, \quad 3, \ldots, a-2, a-1$$

We need at least one number of $1, 2, \ldots, a$ to be written in the first a columns of the winning ticket, so that there must be a good number in the first a columns. To make this work, $1, 2, \ldots, a$ should not be able to be written in the last $50-a$ columns. Thus $50-a < a$, $a > 25$, so $a \geqslant 26$. Therefore $k = 26$ is a winning number.

Next we prove that $k = 25$ is not a winning number. Actually, considering a 25×50 table, we prove that we can write a winning ticket P with suitable numbers $x_1, x_2, \ldots, x_{50}$, such that there is no good number in the table. First write the number 1 in the ticket P. Since there are 25 "1"s in the table and there are 50 columns, there must exist a column which does not contain the number 1. Write 1 in this column in P. Similarly, write the numbers $2, 3, \ldots, a-1$ in turn until a cannot be written. Then $a \geqslant 26$, since otherwise there are only $a-1 \leqslant 24$ numbers in the ticket, occupying 24 blanks in the ticket P; also there are at most 25 as in the table, occupying 25 columns, so there are only $24 + 25$ blanks in P that are occupied, leaving at least one blank fill with number a, whichisa contradiction. Now since there are only 25 as in the table which occupy 25 columns, there will be 25 blanks in the ticket P which can be filled with a. Since a cannot be written down, these 25 blanks must already be filled with numbers (denoted by x_1, x_2, ..., x_{25}) from $\{1, 2, \ldots, a - 1\}$. Consider a blank D in P. The column containing D has 25 numbers, so there are at most 25 numbers which cannot be written in D. Thus there is at least one number from $\{a, x_1, x_2, \ldots, x_{25}\}$ which can be written in D. Since a cannot be written in D, there is at least one number from $\{x_1, x_2, \ldots, x_{25}\}$ which can be written in D. Then we write x_i in D, and write a in the original position of x_i. In this way, the ticket P can be filled so that each number is different from the numbers in the same column.

In conclusion, the minimum value of k is 26.

Example 4. There are $n(n > 3)$ actors in a group. They have made some shows, each of which involves three actors performing. In one occasion they found that it is possible to arrange some shows, so that each pair of actors has exactly one chance to perform in the same show. Find the minimum value of n. (Original)

Solution. Represent the n actors by n points. If a pair of actors performs in the same show, draw a line between the two corresponding points.

Then the requirement is equivalent to the dividing the complete graph K_n into several triangles K_3, such that each edge belongs to only one K_3.

Apparently, $C_3^2 \mid C_n^2$, thus $6 \mid n(n - 1)$, so $3 \mid n$ or $3 \mid (n - 1)$.

Besides (study another property of n to narrow the encirclement), consider the edges containing a point A; there are $n - 1$ edges, and each edge belongs to a K_3. Thus there are $n - 1$ $K_3{}'$s containing point A. Since each K_3 containing point A has two edges containing point A, we know that each K_3 is counted twice. So $2 \mid (n - 1)$, n is odd.

From the above statements, $3 \mid n$ (n is odd) or $6 \mid (n - 1)$. Thus $n = 6k + 3$ or $n = 6k + 1(k \in \mathbf{N}^*)$. Thus $n \geqslant 7$.

When $n = 7$, let the seven points be marked with $0, 1, 2, \ldots, 6$. For $m = 0, 1, 2, \ldots, 6$, let $m, m + 1, m + 3$ form a K_3 (the indices are considered modulo 7), then the 7 $K_3{}'$s satisfy the requirements.

In conclusion, the minimum value of n is 7.

Example 5. If a positive integer n satisfies the following condition: there exists a sequence of n real numbers, where the sum of any 17 consecutive terms is positive, and the sum of any 10 consecutive terms is negative, find the maximum value of n (adapted question).

Solution. The maximum value of n is 25.

First we prove that $n \leqslant 25$ by contradiction.

Assume that there are n real numbers a_1, a_2, ..., a_n, $n \geqslant 26$, that have such property:

(1) The sum of any seven consecutive terms is positive.

Actually, consider the sum of any seven terms $a_{i+1} + a_{i+2} + \cdots + a_{i+7}(i = 0, 1, 2, \ldots, n-7)$. If $i \geqslant 10$, then since $(a_{i-9} + a_{i-8} + \cdots + a_i) + (a_{i+1} + a_{i+2} + \cdots + a_{i+7}) > 0$, and $a_{i-9} + a_{i-8} + \cdots + a_i < 0$, we have $a_{i+1} + a_{i+2} + \cdots + a_{i+7} > 0$.

If $i \leqslant 9$, then $i + 17 \leqslant 9 + 17 = 26 \leqslant n$.

Since $(a_{i+1} + a_{i+2} + \cdots + a_{i+7}) + (a_{i+8} + a_{i+9} + \cdots + a_{i+17}) > 0$, and $a_{i+8} + a_{i+9} + \cdots + a_{i+17} < 0$, we have $a_{i+1} + a_{i+2} + \cdots + a_{i+7} > 0$.

The sum of any three consecutive terms is negative.

Actually, consider the sum of three continuous terms $a_{i+1} + a_{i+2} + a_{i+3}(i = 0, 1, 2, \ldots, n-3)$. If $i \geqslant 7$, then since $(a_{i-6} + a_{i-5} + \cdots + a_i) + (a_{i+1} + a_{i+2} + a_{i+3}) < 0$, and $a_{i-6} + a_{i-5} + \cdots + a_i > 0$, we have $a_{i+1} + a_{i+2} + a_{i+3} < 0$.

If $i \leqslant 6$, then $i + 10 \leqslant 6 + 10 = 16 \leqslant n$.

Since $(a_{i+1} + a_{i+2} + a_{i+3}) + (a_{i+4} + a_{i+5} + \cdots + a_{i+10}) < 0$, and $a_{i+4} + a_{i+5} + \cdots + a_{i+10} > 0$, we have $a_{i+1} + a_{i+2} + a_{i+3} < 0$.

(2) The sum of any four consecutive terms is positive.

Actually, consider the sum of any four consecutive terms $a_{i+1} + a_{i+2} + a_{i+3} + a_{i+4}(i = 0, 1, 2, \ldots, n-4)$. If $i \geqslant 3$, then because $(a_{i-2} + a_{i-1} + a_i) + (a_{i+1} + a_{i+2} + a_{i+3} + a_{i+4}) > 0$ and $a_{i-2} + a_{i-1} + a_i < 0$, we have $a_{i+1} + a_{i+2} + a_{i+3} + a_{i+4} > 0$.

If $i \leqslant 2$, then $i + 7 \leqslant 2 + 7 = 9 \leqslant n$. Since $(a_{i+1} + a_{i+2} + a_{i+3} + a_{i+4}) + (a_{i+5} + a_{i+6} + a_{i+7}) > 0$, and $a_{i+5} + a_{i+6} + a_{i+7} < 0$, we have

$$a_{i+1} + a_{i+2} + a_{i+3} + a_{i+4} > 0.$$

(3) Each term is positive.

Actually, for any term $a_i (i = 1, 2, \ldots, n)$, if $i \geqslant 4$, then because $(a_{i-3} + a_{i-2} + a_{i-1}) + a_i > 0$, and $a_{i-3} + a_{i-2} + a_{i-1} < 0$, we have $a_i > 0$.

If $i \leqslant 3$, then $i + 3 \leqslant 3 + 3 = 6 \leqslant n$. Since $a_i + (a_{i+1} + a_{i+2} + a_{i+3}) > 0$, and $a_{i+1} + a_{i+2} + a_{i+3} < 0$, thus $a_i > 0$.

Apparently, (4) contradicts the condition that the sum of any 10 consecutive terms is negative, so $n \leqslant 25$.

Besides, when $n = 25$, we can construct the 25 real numbers satisfying the conditions.

Let these 25 real numbers be $a_1, a_2, \ldots, a_{25}$, we first study the necessary conditions which they should satisfy. From the above statements, for $i = 0, 1, 2, \ldots, 25 - 7$ and except for $i = 9$, the sum $a_{i+1} + a_{i+2} + \cdots + a_{i+7}$ should be positive. Then, except for $i = 6, 16$, the sum $a_{i+1} + a_{i+2} + a_{i+3}$ should be negative. To simplify the construction, we shall use symmetry, periodicity, and some parameters. Assume that these 25 real numbers are: $a, a, b, a, a, b, c, c, c, b, a, a, d, a, a, b, c, c, c, b, a, a, b, a, a$, where a, b, c, d are some parameters to be determined, which satisfy $a + a + b < 0$. Choose $a = 1$, $b = -3$, then the sequence turns into

$$1, 1, -3, 1, 1, -3, c, c, c, -3, 1, 1, d, 1, 1, -3,$$
$$c, c, c, -3, 1, 1, -3, 1, 1.$$

To make the sum of any 10 consecutive terms negative, it is required that $d + (1 + 1 - 3) + 3c + (-3 + 1 + 1) < 0$, thus $d + 3c < 2$.

To make the sum of any 17 consecutive terms positive, it is required that $(1 - 3) + 3c + (-3 + 1 + 1) + d + (1 + 1 - 3) + 3c + (-3 + 1) > 0$, thus $d + 6c > 6$. Therefore we have $6 - 6c < d < 2 - 3c$, so $c > \frac{4}{3}$. Let $c = 1.5$, then $-3 < d < -2.5$, let $d = -2.6$. We have the following sequence:

$$1, 1, -3, 1, 1, -3, 1.5, 1.5, 1.5, -3, 1, 1, -2.6, 1, 1,$$
$$-3, 1.5, 1.5, 1.5, -3, 1, 1, -3, 1, 1.$$

To obtain a sequence including integers of small absolute values, let $d = -2.75$, the sequence turns into:

$$1, 1, -3, 1, 1, -3, 1.5, 1.5, 1.5, -3, 1, 1, -2.75, 1, 1,$$
$$-3, 1.5, 1.5, 1.5, -3, 1, 1, -3, 1, 1.$$

Multiplying each term by 4, we get

$$4, 4, -12, 4, 4, -12, 6, 6, 6, -12, 4, 4, -11, 4, 4,$$
$$-12, 6, 6, 6, -12, 4, 4, -12, 4, 4.$$

Note that in the above construction, let $a = c$, $b = d$, then the corresponding sequence is: $a, a, b, a, a, b, a, a, a, b, a, a, b, a, a, b, a, a, a, b, a, a, b, a, a$, where a, b are parameters to be determined, which satisfy $a + a + b < 0$, $2(a + a + b) + 3a + b < 0$, and $6(a + a + b) - b > 0$. Let $a = 8$, $b = -19$, we have the following construction:

$$8, 8, -19, 8, 8, -19, 8, 8, 8, -19, 8, 8, -19, 8, 8,$$
$$-19, 8, 8, 8, -19, 8, 8, -19, 8, 8.$$

Exercise 12

(1) Assume that A is a k-element subset of set $\{1, 2, 3, \ldots, 16\}$, and that the sums of any two subsets of A are not equal. For any $k + 1$-element subset B of set $(1, 2, 3, \ldots, 16)$ which is a superset of A, there exist two subsets of B whose sums are equal. (i) Prove that $k \leqslant 5$; (ii) find the maximum and minimum values of the sum of members of A. (2002 Bulgaria Winter Mathematical Olympiad)

(2) There are 2000 councilors in the council, and 200 expenditures in the budget to be examined. Each councilor will prepare a budget proposal and list the amounts of the expenditures, of which the sum

does not exceed S. During the council, the amount of each expenditure is chosen to be the maximal amount such that it is no more than the amounts proposed by at least k councilors. What is the smallest k so that it is guaranteed that the sum of the approved amounts does not exceed S? (The 54th Moscow Mathematical Olympiad)

(3) There are $n(n \geqslant 5)$ football teams in the league. Any two teams play one exactly game. The rule is such that a win gives three points, a draw gives one point and a loss gives zero point. After the league ends, some teams may be disqualified and their match results will be canceled. Among the rest of the teams, the team with the highest score (strictly higher than any other teams) will be the champion (if there is only one team that is not disqualified, then it will be the champion). Denote by f_i the minimal number of teams needed to be disqualified in order to make the ith team the champion. Find the maximum and minimum value of $F = f_1 + \cdots + f_n$. (The 53rd Belarus Mathematical Olympiad)

Chapter *13* Considering Special Cases

Considering special case means to study some special cases of the problem which are relatively simpler than the general situation, and try to find the approach to the solution from these special cases. It usually includes four cases.

Case 1. Study the "*worst*" case, which is one of the most special cases.

Case 2. Study the special cases suggested by the sufficient or necessary conditions. For example, in order to find a sufficient condition for property P, we may first assume that this property is violated, and then deduce several other properties from this assumption. Any assumption that violates one of the new properties is now a good sufficient condition for P.

Case 3. First study the problem in special cases, and then reduce the general case to these special cases.

Case 4. Find general rules from special cases and then guess the conclusion of the problem, then prove the conclusion by induction.

Example 1. There are n middle schools in a city. The ith middle school sends c_i students ($1 \leqslant c_i \leqslant 39$) to watch a football game in a stadium, where $\sum_{i=1}^{n} c_i = 1990$. There are 199 seats in each row of the stand. It is required that the students in the same school sit in the same row. At least how many rows should there be, so that this is always possible?

Analysis and Solution. First we consider the worst case. The worst case means that there are as many empty seats in each row as possible. Apparently, it is easier to arrange the seats when the numbers of students from the schools are very different, because we can arrange that the schools with smaller numbers of students fill the empty seats. So the worse case is that the numbers of students from the schools are large and close to each other. Therefore, we assume that the numbers of students from the schools are equal. Assuming that each school has r students, we consider the value of r that makes the number of empty seats largest. The estimation is listed below in the table:

Student number r of each school	39	38	37	36	35	34	33	32	31	30	29	$r \leqslant 28$
Number of school in each row	5	5	5	5	5	5	6	6	6	6	6	...
Number of empty seats in each row	4	9	14	19	24	29	1	7	13	19	25	$t \leqslant 28$

From the above table, we can see that when each school has 34 students, the number of empty seats in each row is 29, which is the largest. Noticing that $1990 = 34 \times 58 + 18$, we then consider the situation when there are 34 students in each of 58 schools, and 18 students in another school. In this case, each row can hold at most six schools, and there can be only one row holding six schools: the row where the school with 18 students is located.

Thus, we need at least $1 + \left\lceil \dfrac{58 - 6}{5} \right\rceil + 1 = 12$ rows.

Finally, we prove that 12 rows are sufficient in any case. A natural arrangement is as follows.

First we fill the first row to minimize the number of empty seats x_1 in the first row. Then we fill the second row, again minimizing the number x_2 of empty seats in the second row. It is obvious that $x_1 \leqslant x_2$, since otherwise we could fill in the first row with the students in the second row, which contradict the assumption that x_1 is minimized. Continue this process until the 11th row is filled. Now we prove that with this arrangement, the rest of the students can be put in the

12th row.

Actually, assume that there are x students not seated after the 12 rows are filled. Then $x > x_{12}$, since otherwise these x students can be arranged in the 12th row. Thus $x = 1990 - \sum_{i=1}^{12} (199 - x_i) > x_{12}$, thus $x_1 + x_2 + \cdots + x_{11} > 398$. So $398 < x_1 + x_2 + \cdots + x_{11} \leqslant 11x_{11}$, thus $x_{11} \geqslant 37$. Therefore the number of students in each school in the 12nd row is no less than 38 (note that students in any school in the 12nd row cannot be arranged in the x_{11} empty seats of the previous row). Since each school has at most 39 students, the number of students of each school in the 12nd row is either 38 or 39. Since $5 \times 39 < 199 < 6 \times 38$, there must be five schools in the 12nd row. Assume that there are 38 students in k schools, 39 students in $5 - k$ schools. Then the number of empty seats in the 12nd row will be

$$x_{12} = 199 - k \times 38 - (5 - k) \times 39 = 199 + k - 5 \times 39 = 4 + k \leqslant 9.$$

However, $x_{11} \geqslant 37 > x_{12}$, which contradicts the assumption that x_{11} is minimized.

Note. There is a more intuitive arrangement. First arrange the first 10 rows. Arrange the schools to each row until no more school can be arranged, then arrange the next row, and so on, until 10 rows are filled. Now estimate the number of schools that are not arranged. It is easy to know that there are at most nine schools not arranged, otherwise there are 10 schools not arranged, each of which cannot be arranged in the first 10 rows, which means that the number of students from each of the remaining schools plus the number of students in any of the first 10 rows is more than 199. Thus the total number is more than 1990, which is a contradiction. Therefore there are at most 10 schools that are not arranged. Since each row can accommodate at least five schools ($5 \times 39 < 199$), two rows will suffice for the remaining schools.

Example 2. There are r stones in an $n \times n$ chessboard. Each cell has at most one stone. Suppose that the r stones have the following property

P : each row and each column of the chessboard has at least one stone, and if any stone is removed, the property P will not hold. Find the maximum value r_n of r.

Analysis and Solution. First consider a simple case. To simplify the statement, we call a stone *"removable"* if property P holds after the stone is removed from the chessboard.

When $n = 2$, we have $r_2 < 3$; otherwise when three stones are put on the chessboard, there must be a stone at the corner which is removable, which is a contradiction.

When $n = 3$, we have $r_3 < 5$; otherwise, when five stones are put on the chessboard, there must be a row containing two stones. Assume that the first row has two stones in the first two cells (a_{11}, a_{12}). Then, the first and second columns contain no more stones (otherwise assume that there is one more stone in the first column, then a_{11} will be removable), so the rest three stones lie in the 3rd column. Thus the stone a_{13} in the first cell of the 3rd column is removable.

In general, we guess that $r_n < 2n - 1$ for the positive integer n.

Actually, from the above statements we can see that, if a row has two stones a, b, then the column which contains a or b has no more stones. For example, if the column containing a has one more stone, then a is removable, which is a contradiction. With this property, we can remove a row and a column from the chessboard to reduce the problem with n to the problem with $n - 1$.   ($*$)

First proof. Induct on n.

Assume that ($*$) holds for positive integers less than n. Consider an $n \times n$ chessboard. To apply the assumption, we should remove one row and one column from the chessboard, so that there are at least $2n - 3$ stones left on the chessboard. Thus we should find a column and a row that has only one stone. However, the number of stones on the chessboard is no less than $2n - 1$, not precisely $2n - 1$. So the pigeonhole principle cannot be applied to find a row and a column with only one stone. Nevertheless, note that we have proved previously that: if some row has two stones, then the two columns at which the two stones are located have only one stone. Thus, we can find the column satisfying the condition from this property.

Since there are at least $2n - 1$ stones in the board, by the pigeonhole principle, there is at least one row having two stones. Assume that the first row has two stones a, b located in the first two cells of this row. Then the first and second columns cannot have any more stones. If the first column has one more stone, then a is removable. If the second column has one more stone, then b is removable. Thus the rest $2n - 3$ stones lie in the last $n - 2$ columns. By the pigeonhole principle, there is a column which contains two stones. Assume that the ith grid and jth grid of the third column have stones c and d respectively, then one of i, j is not equal to 1. Assume that $i \neq 1$, then the row containing c cannot have more stones, otherwise c is removable. Thus, the column containing a has only one stone, and that the row containing c has only one stone Figure 13.1. Removing this column and this row, and applying the inductive hypothesis to the rest of the chessboard, we can prove ($*$).

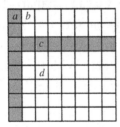   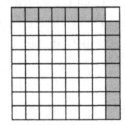

Figure 13. 1   Figure 13. 2

Finally, by Figure 13.2, $r = 2n - 2$ is possible. Thus $r_n = 2n - 2$.

Second proof. To find a removable stone, we consider a necessary condition under which a stone is removable.

If A is removable, then the row and column of A should have at least two stones. So we want to find a row with two stones, and then find a column with two stones.

Assume that the number of stones in the ith row is a_i $(i = 1, 2, \ldots, n)$, assume that $a_1 \leqslant a_2 \leqslant \cdots \leqslant a_n$, and $a_1 + a_2 + \cdots + a_n = 2n - 1$.

Since there is at least one stone in each row, we know $a_1 = 1$.

Assume that $a_1 = a_2 = \cdots = a_i = 1(1 \leqslant i \leqslant n-1)$, $2 \leqslant a_{i+1} \leqslant a_{i+2} \leqslant \cdots \leqslant a_n$, then $a_{i+1} + a_{i+2} + \cdots + a_n = 2n - 1 - (a_1 + a_2 + \cdots + a_i) = 2n - 1 - i = n + (n - 1 - i) \geqslant n$.

The i stones in the first i rows occupy at most i columns (as shown in the shadow grids of Figure 13.3). Assume that these i stones are located in the first $k(k \leqslant i)$ columns. If there is stone A lying in the last $n - i$ rows and the first k columns, then since the row containing A has at least two stones, A will be removable. If the stones in the last $n - i$ rows all lie in the last $n - k$

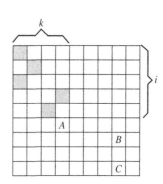

Figure 13. 3

columns, then since $n - k < n$, and $a_{i+1} + a_{i+2} + \cdots + a_n \geqslant n$, there must be a column with two stones B, C. But the row containing B has two stones, so B is removable.

Example 3. How many stones can be put on a 19×89 chessboard, with any 2×2 square in the chessboard containing no more than two stones?
Analysis and solution. Note that 19 and 89 are odd; we will consider a general $(2m - 1) \times (2n - 1)$ chessboard (at least one of m, n should be larger than 1). We may try to induct on m. We assume that the maximum number of stones on the chessboard is r_m.

(1) When $m = 1$, each grid can be filled with a stone. So $r_1 = 2n - 1$.

(2) When $m = 2$, with the pattern from Figure 13.4, we can guess that $r_2 = 2(2n - 1)$, thus the $3 \times (2n - 1)$ chessboard can have at most $4n - 2$ stones.

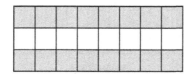

Figure 13. 4

Two rows need to be removed to convert a $3 \times (2n-1)$ chessboard into a $1 \times (2n-1)$ chessboard. How many stones can be put on the removed rows? To answer this, it is helpful to study the $2 \times (2n-1)$ chessboard, which is a situation we skipped. So, we go back to the $2 \times (2n-1)$ case.

From Figure 13.5, in a $2 \times (2n-1)$ chessboard, we have $r \leqslant 2n$, and equality holds only in the situation shown in Figure 13.5.

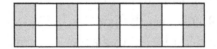

Figure 13.5

Actually, the first column has at most two stones; the rest $2n-2$ columns can be divided into $n-1$ "2×2" boxes, where each box has at most two stones. So $r \leqslant 2 + 2(n-1) = 2n$. If $r = 2n$, then the first column must have two stones, and each 2×2 box has exactly two stones. Thus, the first column has two stones, the second column has no stone, the third column has two stones, and so on. Thus each of the odd columns has two stones and each of the even columns has no stone. Therefore equality holds in a unique situation.

Now consider the $3 \times (2n-1)$ chessboard. We need to prove that $r \leqslant 2(2n-1)$. Naturally, we divide the $3 \times (2n-1)$ chessboard into a $2 \times (2n-1)$ chessboard (A) and a $1 \times (2n-1)$ chessboard (B).

From the previous discussion, we have $r_A \leqslant 2n$, $r_B \leqslant 2n-1$. Then,

$$r = r_A + r_B \leqslant 4n - 1. \qquad \textcircled{1}$$

If $\textcircled{1}$ holds, then $r_A = 2n$, $r_B = 2n-1$. Then pattern of the stones on the chessboard is shown in Figure 13.6. However, there is a 2×2

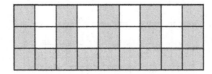

Figure 13.6

box containing three stones, which is a contradiction.

In general, for any $(2m-1) \times (2n-1)$ chessboard, if $m \leqslant n$, we prove that $r \leqslant m(2n-1)$.

Induct on m. When $m = 1$, the result holds trivially. Assume that the conclusion holds for positive integers smaller than m. Consider a $(2m-1) \times (2n-1)$ chessboard. We divide it into a $2 \times (2n-1)$ chessboard A and a $(2m-3) \times (2n-1)$ chessboard B. From the previous assumptions and discussions we know that $r_A \leqslant 2n$, $r_B \leqslant (m-1)(2n-1)$, thus

$$r = r_A + r_B \leqslant 2n + (m-1)(2n-1) = m(2n-1) + 1. \qquad ②$$

If the equality in ② holds, then $r_A = 2n$, $r_B = (m-1)(2n-1)$. $r_A = 2n$ indicates that the whole $(2m-1) \times (2n-1)$ chessboard contains n stones in the first row. Thus, divide the last $2m-2$ rows into $m-1$ "$2 \times (2n-1)$" chessboards, where each $2 \times (2n-1)$ chessboard contains no more than $2n$ stones. Then the number of stones is $r \leqslant n + 2n(m-1) = 2mn - n \leqslant 2mn - m = m(2n-1)$, which contradicts the fact that $r = m(2n-1) + 1$. So equality in ② does not hold, and hence $r \leqslant m(2n-1)$. This completes the proof.

Finally, put a stone in each grid of the odd rows of the chessboard, then $r = m(2n-1)$. So $r_m = m(2n-1)$. In particular, when $m = 10$, $n = 45$, the 19×89 chessboard can have a maximum of 890 stones.

Example 4. There is a 9×9 chessboard with grids colored black and white. The grids next to each white grid include more black grids than white grids, and the grids next to each white grid include more white grids than black grids (a grid is next to another when these two grids share an edge). Find the maximum value of the difference between the numbers of black and white grids for all possible painting patterns. (The 53rd Belorussia Mathematical Olympiad)

Analysis and Solution. To satisfy the requirements, each grid has at most one neighbor with the same color. Thus the following specific situations are impossible: (1) three grids in an L pattern having the same color. (2) three grids in a 1×3 rectangle having the same color.

If any two grids next to each other in the chessboard have different colors, then the difference between the numbers of black and white grids is no more than 1.

If the chessboard has two grids (A, B) neighboring each other with the same color, assume that A and B lie in the same row (as Figure 13.7 shows). Consider the next row to the row containing A, B. Since the chessboard has no L pattern with the same color, the grids next to A, B in this row must the same color. Thus we know that, in the two columns containing A and B, grids in each row have the same color, and that the grids in the each column are alternately black and white (any two grids next to each other having different colors). Next we prove that, in any column in the chessboard, there cannot be a pair of neighboring grids, which have the same color. Otherwise, assume that (P, Q) is a pair of neighboring grids in a column with the same color. As in the previous arguments, in the two rows containing P and Q respectively, grids in each column have the same color, and that the grids in each row are alternately black and white. Then we look at the four grids lying in the intersection points of the two rows A, B and two columns C, D. From the above arguments, we know that A, B, C, D must have the same color, which is a contradiction. Thus, the grids in each column of the chessboard are alternately black and white.

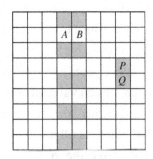

Figure 13.7

Removing the first row; we see that each column of the rest of the chessboard has the same number of black and white grids. Thus the numbers of black and white grids in the rest of the chessboard are equal. In the first row, since there are no 1×3 rectangle of the same color, in any three consecutive grids the difference between the numbers of black and white grids is no more than 1. Thus the difference between the numbers of black and white grids in the whole chessboard is no more than 3.

Thus, for any pattern satisfying the requirements, the difference between the numbers of black and white grids is no more than 3. Also Figure 13.8 shows that there is a pattern with difference 3.

In conclusion, the maximum value is 3.

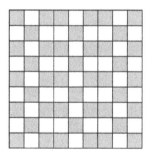

Figure 13.8

Example 5. Fix a positive integer $n \geqslant 3$, let a_1, a_2, ..., a_n be n different real numbers, whose sum is positive. If a permutation of these numbers is such that for any $k = 1, 2, \ldots, n$, $b_1 + b_2 + \cdots + b_k > 0$, then the permutation is called "*good*". Find the minimum number of "*good*" permutations. (2002 Bulgaria National Mathematical Olympiad in Regional Level)

Analysis and Solution. Consider the worst case. We need that the number of negative terms in a_1, a_2, ..., a_n to be as many as possible in order to make the condition $b_1 + b_2 + \cdots + b_k > 0$ hard to satisfy. Let a_2, a_3, ..., a_n be negative, and $a_1 = 1 - a_2 - \cdots - a_n$. Then for any good permutation $(b_1, b_2, \ldots, b_n)$, $b_1 = a_1$. Also $(b_2, \ldots, b_n)$ can be any permutations of $a_2, \ldots, a_n$, so there are $(n-1)!$ permutations satisfying this condition.

Next we prove that there are at least $(n-1)!$ good permutations. Note that $(n-1)!$ is the number of different cyclic permutations of a_1, $a_2, \ldots, a_n$. We first prove that each cyclic permutation corresponds to at least one good permutation. To simplify the statements, we call the first term of a good permutation a "*good term*". We only have to prove that each cyclic permutation has at least a good term.

First proof. Induct on n. When $n = 1$, the conclusion apparently holds. Assume that the conclusion holds for any $n < k$, and consider the case of $n = k$. If all numbers are positive, then the conclusion apparently holds because each number is a good term. If there is at least one nonpositive number, since $a_1 + a_2 + \cdots + a_n > 0$, then there must be at least one positive number. For each positive number, group it and all the numbers between it and the next positive number anti-clockwise, then each group has at least one number, and there is at least one group having at least two numbers (since there is at least one nonpositive number), so there are at most $k-1$ groups. Summing up each group, we get less than k sums. Arrange these sums in the circle in the order of their corresponding groups; since the sum of these k sums is positive, by induction hypothesis, there is a sum that is a good term. Then, the positive number in the corresponding group of the "*good term*" sum is a good term with respect to the whole circle. The conclusion is proved.

Second proof. Use the extremal principle. For any cyclic permutation $(b_1, b_2, \ldots, b_n)$, consider the sums of the numbers starting from b_i: $b_i + b_{i+1} + \cdots + b_{i+t}$, where for subscripts larger than n we take modulo n. For all $i = 1, 2, \ldots, n$ and all $t = 0, 1, 2, \ldots, n-1$, there must exist a minimal sum $b_i + b_{i+1} + \cdots + b_{i+t}$. Since there is at least one nonpositive number, we have $b_i + b_{i+1} + \cdots + b_{i+t} \leqslant 0$. In all of the minimal sums, we choose the sum with the largest number of terms $t + 1$, and prove that b_{i+t+1} is a good term.

Actually, if there is a positive integer k such that $b_{i+t+1} + b_{i+t+2} + \cdots +$

$b_{i+t+k} \leqslant 0$, then $(b_i + b_{i+1} + \cdots + b_{i+t}) + (b_{i+t+1} + b_{i+t+2} + \cdots + b_{i+t+k}) \leqslant b_i + b_{i+1} + \cdots + b_{i+t}$, which contradicts the fact that $b_i + b_{i+1} + \cdots + b_{i+t}$ is minimal and has the largest number of terms.

Since there are at least $(n-1)!$ cyclic permutations, and each cyclic permutation corresponds to at least one good permutation, and different cyclic permutations correspond to different good permutations, there are at least $(n-1)!$ good permutations.

Example 6. There are n native residents in an island. Each two of them are either friends or enemies of each other. One day, the leader requires each resident (including himself) to make a rock necklace according to the following principles: the necklaces of any two friends should have at least one rock in common, and the necklaces of any two enemies should have no rock in common (it is allowed that there is no rock in a necklace). Prove that: to carry out the order of the leader, $\left[\dfrac{n^2}{4}\right]$ different kinds of rocks are necessary and sufficient. If the number of types of rocks is less than $\left[\dfrac{n^2}{4}\right]$, the command may not be possible to be carried out. (2002 Croatia National Mathematical Olympiad)

Analysis and Solution. When $n = 1$, the conclusion apparently holds. Assume $n > 1$, and the minimum number of types of rocks required is S_n. When $n = 2$, assume that the two people are A, B. If A and B are enemies, then $S_2 = 0$; if A and B are friends, then $S_2 = 1$, so the conclusion holds. When $n = 3$, assume that the three people are A, B, C. If A, B, C are enemies to each other, then $S_3 = 0$; if A, B, C are friends to each other, then $S_3 = 1$. If in A, B, C there are two pairs of enemies and one pair of friends, then the pair of friends need a rock, and $S_3 = 1$. If there are two pairs of friends and a pair of enemy, then two different types of rocks are required from the two pairs of friends, since otherwise, the three people would have the same set of rocks, but two of them are enemies, which is a contradiction. Thus, $S_3 = 2$.

From the previous special cases, we found a useful rule: if among three people there is one enemy pair and two friend pairs, then the two friend pairs need two different types of rocks.

Consider $n = 4$. Assume that there are four people A, B, C, D. If there are no more than four friend pairs, then $S_4 \leqslant 4$. If there are five friend pairs, and the other pair is an enemy pair, assume that A, B are enemies. Then ACD is a friend triangle, and assume they have rocks with the same type 1. BCD is a friend triangle, and assume that they have rocks with the same type 2. Then, A, B, C, D have necklaces $\{1\}$, $\{2\}$, $\{1, 2\}$, $\{1, 2\}$ respectively, which satisfy the requirements, so $S_4 = 2$. If A, B, C, D are friends with each other, then $S_4 = 1$.

Now consider the case when $S_4 = 4$. There should be four friend pairs, and the rest two are enemy pairs. If the two enemy pairs share a common person, assume that A, B are enemies and A, C are enemies. Since BCD is a friend triangle, assume they have rock 1; since AD are friends, assume they have rock 2. Then the necklaces of A, B, C, D are $\{2\}$, $\{1\}$, $\{1\}$, $\{1, 2\}$, which satisfies the requirements. In this case $S_4 = 2$. If two enemy pairs have no common person, we assume that A, B are enemies and C, D are enemies. Then A, B, C, D are divided into two groups, each group having two people, and that the two people in the same group are enemies and any two people from different groups are friends. Now, each friend pair corresponds to a common rock. We prove that the four friend pairs correspond to four different rocks. Actually, if there are two friend pairs sharing the same rock, the two friend pairs include at least three people who have the same rock, and there must be two people in the same group among these three people, which contradicts the fact that the people in the same group are enemies. So $S_4 = 4$.

From the above special cases, it is easy to complete the construction in the general situation. When n is odd, let $n = 2k + 1$. Divide the n people into two groups with k and $k + 1$ people respectively. Each pair of people in the same group is an enemy pair, and each pair of people in different groups is a friend pair. Now we have $k(k + 1)$ friend pairs.

We prove that: the $k(k+1)$ friend pairs correspond to different rocks. In fact, if there are two friend pairs having the same rock, since these two friend pairs include at least three people, we have three people sharing the same rock. Since at least two people will fall into the same group, this contradicts the fact that people from the same group are enemies. So we need at least $k(k+1) = \left\lceil \dfrac{n^2}{4} \right\rceil$ rocks. When n is even, let $n = 2k$. Similarly, dividing n people into two groups with k people, we see that $k^2 = \left\lceil \dfrac{n^2}{4} \right\rceil$ rocks are necessary.

Next we prove that when $S = \left\lceil \dfrac{n^2}{4} \right\rceil$, the residents can make necklaces satisfying the requirements. We induct on n.

Assume that the conclusion holds when $n = k$. Considering $n = k + 1$, we analyze the increment $\triangle = S_{k+1} - S_k = \left\lceil \dfrac{(k+1)^2}{4} \right\rceil - \left\lceil \dfrac{k^2}{4} \right\rceil$, to calculate $\triangle$, we consider the parity of k.

When k is odd, let $k = 2r + 1$, then

$$\triangle = \left\lceil \frac{(k+1)^2}{4} \right\rceil - \left\lceil \frac{k^2}{4} \right\rceil = (r+1)^2 - (r^2 + r) = r + 1 = \frac{k+1}{2}.$$

When k is even, let $k = 2r$, then

$$\triangle = \left\lceil \frac{(k+1)^2}{4} \right\rceil - \left\lceil \frac{k^2}{4} \right\rceil = (r^2 + r) - r^2 = r = \frac{k}{2}.$$

From this we can see that from k to $k + 1$, the number of rocks required has increased by $\dfrac{k+1}{2}$ (when k is odd) or $\dfrac{k}{2}$ (when k is even). This inspires the idea to divide the k people to $\dfrac{k+1}{2}$ (when k is odd) or $\dfrac{k}{2}$ (when k is even) groups, each group having at most two people (when k is odd then there is one group with only one person, and the rest groups have two people; when k is even all groups have two people). We hope that when a new person P is added, P and each group need at most one new rock. Is it possible? If the two people in some group are both enemies to P, then no new rocks are needed. If in some group one person is an enemy of P and the other person is a friend of P, then one new rock is needed for the friend and P. However, if the two people in a group are enemies of each other, and both are friends to P, then each person needs one new rock, so two new rocks are needed. Therefore we can see that, people who are friends of P and enemies to each other cannot be grouped together. Unfortunately, such grouping may not be possible because P may have more enemies than friends. Now, consider another point of view; if we fix two people A, B who are friends, then at most one rock is needed between each of the other people and A, B. So the conclusion can be proved by moving from k to $k + 2$.

Assume that the conclusion holds when $n = k$, consider the case when $n = k + 2$.

If there are no friends between these $k + 2$ people, the conclusion holds apparently (no rock is needed). Otherwise, assume that A, B are friends, then from the assumption, at most $\left\lceil \dfrac{k^2}{4} \right\rceil$ rocks are needed among the other k people. For any person P in these k people, P and A, B form a triangle. We prove that at most one new rock is needed in this triangle. If one of A, B is enemy to P, then a new rock is needed between the pair of friends. If both A, B and P are friends, one new rock is needed for all of them. Because P is arbitrary, we need at most k new rocks for k people, and at most one new rock is needed between A and B. So at most $\left\lceil \dfrac{k^2}{4} \right\rceil + k + 1 = \left\lceil \dfrac{(k+2)^2}{4} \right\rceil$ new rocks are needed for $k + 2$ people.

From the above statements, the conclusion is proved.

Example 7. Assume that m, n are positive integers, $m < 2001$, $n < 2002$. There are 2001×2002 different real numbers, and these numbers are filled in the cells of a 2001×2002 chessboard, where each cell is filled with one number. If there is a cell with the number smaller than at least m numbers in the same column and at least n numbers in the same row, then the cell is called a "*bad*" cell. For all possible arrangements, find the minimum value of the number S of bad cells. (2002 Vietnam Mathematical Olympiad)

1	2	⋯	2002
2003	2004	⋯	4004
⋮	⋮	⋮	⋮
$2000 \times 2002 + 1$	$2000 \times 2002 + 2$	⋯	2001×2002

We guess that the minimum value of S is $(2001 - m)(2002 - n)$. In general, for a $p \times q$ ($m < p$, $n < q$) chessboard, the minimum value of S is $(p - m)(q - n)$. We prove this by induction.

Assume that the result holds for $p \times q$ chessboards. Consider a $(p + 1) \times q$ chessboard. To use the assumption, we should remove a row, which has at least $(p + 1 - m)(q - n) - (p - m)(q - n) = q - n$ bad cells. But a row has at most $q - n$ bad cells since only the first $q - n$ numbers (from small to big) are less than at least n numbers in this row. For simplicity, if the number in a cell is smaller than at least n numbers in the same row, then the cell is called a "*row-bad*" cell, otherwise it is called a "*row-good*" cell. Similarly we have "*column-bad*" and "*column-good*" cells. Therefore, a bad cell is a "*row-bad*" and "*column-bad*" cell. From the above analysis, we need a row which contains no "*row-bad*" and "*column-good*" cells. By symmetry, we also just need a column which contains no "*column-bad*" and "*row-good*" cells.

Lemma. A $p \times q$ ($m < p$, $n < q$) chessboard either has a row which contains no "*row-bad*" and "*column-good*" cells, or has a column which contains no "*columncolumn-bad*" and "*rowcolumn-good*" cells.

Proof. If the chessboard has no "*row-bad*" and "*column-good*" cells (or has no "*row-good*" and "*column-bad*" cells), then each row (or column) satisfies the requirement. If there are "*row-bad*" and "*column-good*" cells and "*row-good*" and "*column-bad*" cells, we may find the cell A which has the minimal number among these cells. Assume that A is "*row-good*" and "*column-bad*" and has number x. If there is a "*row-bad*" and "*column-good*" cell B in the same row of A, assume that B has number y. On the one hand, since x is minimal, $x < y$; on the other hand, since A is row-good and B is row-bad, $x > y$, which is a contradiction. The lemma is proved. $\square$

Next we prove that: for any $p \times q \, (m < p, \, n < q)$ chessboard, the number of bad cells is no less than $(p - m)(q - n)$.

Induct on $p + q$. When $p + q = m + n + 2$, since $p \geqslant m + 1$, $q \geqslant n + 1$, then $p = m + 1$, $q = n + 1$. Since the cell having the minimum number must be a bad cell, there are at least $1 = (p - m)(q - n)$ bad cells. The conclusion holds. Assume the conclusion holds when $p + q = t$. Consider the case when $p + q = t + 1$. From the lemma, assume that there is a row which does not contain any row-bad and column-good cell. We remove this row, so that the chessboard becomes $(p - 1) \times q$. Since $p + q - 1 = t$, by induction hypothesis, there are no less than $(p - 1 - m)(q - n)$ bad cells. Now we put back the removed row, and the previous bad cells are still bad cells. Also, the row-bad cells in the removed row must be column-bad cells (since the row has no row-bad and column-good cell), and thus are bad cells. Since there are $q - n$ row-bad cells in the removed row, the total number of bad cells are no less than $(p - 1 - m)(q - n) + (q - n) - (p - m)(q \quad n)$. The statement is proved.

Coming back to the original question, we know that the number of bad cells is no less than $(2001 - m)(2002 - n)$. From the previous construction we can see that the minimum number of bad cells is $(2001 - m)(2002 - n)$.

Example 8. A diagonal of a 2006-sided regular polygon is called "*good*" if the two end-points of it divide the boundary of P into two parts, and that each part contains an odd number of sides. In particular, the sides of P are also called good.

Assume that P is divided into triangles with 2003 diagonals which do not intersect each other, find the maximum number of isosceles triangles, two of whose edges are good. (The 47th IMO in 2006)

Solution. We call an isosceles triangle "*good*" if two of its sides are good. First we consider special cases.

For a square, there is only one way of division, and we have two good triangles.

For a regular hexagon, there are essentially three ways of division, and the maximum number of the good triangles is 3. Also we find that when this number reaches maximum, the legs of the good triangles are sides of P.

In general, it is not hard to see that a $2n$-sided regular polygon has at most n good triangles.

Actually, for any triangle ABC in the division, the sides of P are divided into three parts by A, B, C. Denote the number of sides in $A - B$ part of P by $m(AB)$, and so on.

Since $m(AB) + m(BC) + m(CA) = 2006$, if an isosceles triangle has two good sides, then they only have two good sides, and these good sides are legs (otherwise all three sides are good sides, which is impossible).

Consider any good triangle ABC with $AB = AC$. If there is any other good triangle in part AB, then remove all sides of P "*cut*" by the legs of the good triangle; we know that the number of removed sides is even. The process goes on until there is no good triangle in AB part. Since the AB part has an odd number of sides, at least one side α is not removed (if AB is a side of P, then $\alpha = AB$), so α does not belong to legs smaller than AB. Similarly, remove the good triangles in AC pair, and there is one side β which does not belong to legs smaller than AC. Pair $\triangle ABC$ with the set $\{\alpha, \beta\}$.

For two different good triangles $\triangle ABC$, $\triangle A_1B_1C_1$ in a same division, assume that the corresponding sets are $\{\alpha, \beta\}$, $\{\alpha_1, \beta_1\}$ respectively. If $\triangle A_1B_1C_1$ is not in a "*leg part*" of $\triangle ABC$, then $\triangle A_1B_1C_1$ lies in the "*base part*" of the triangle $\triangle ABC$. In this case, the sides in $\{\alpha, \beta\}$ lie in the leg parts of $\triangle ABC$, and the sides of

$\{\alpha_1, \beta_1\}$ lie in BC part of $\triangle ABC$, thus $\{\alpha, \beta\}$, $\{\alpha_1, \beta_1\}$ have no common sides. If $\triangle A_1B_1C_1$ is in a *"leg part"* of $\triangle ABC$, suppose it lies in the AB part. Then the sides in $\{\alpha, \beta\}$ belong to the remaining set of sides of the polygon after removing the sides *"cut off"* by the legs of $\triangle A_1B_1C_1$. Therefore the sets $\{\alpha, \beta\}$, $\{\alpha_1, \beta_1\}$ have no common sides.

Note that there are at most $\dfrac{2006}{2} = 1003$ two-element subsets which do not share any common elements in the 2006 sides, we know that there are at most 1003 good triangles.

Finally, suppose that $P = A_1A_2 \ldots A_{2006}$, then using diagonals $A_1A_{2k+1} (1 \leqslant k \leqslant 1002)$ and $A_{2k+1}A_{2k+3} (1 \leqslant k \leqslant 1001)$ to divide the polygon, we obtain 1003 good triangles.

So the maximum number of good triangles is 1003.

Example 9. Given a positive integer a, let $X = \{a_1, a_2, a_3, \ldots, a_n\}$ be a set of positive integers, where $a_1 \leqslant a_2 \leqslant a_3 \leqslant \cdots \leqslant a_n$. If for any integer p $(1 \leqslant p \leqslant a)$, there is a subset of X such that $S(A) = p$, where $S(A)$ is the sum of elements in set A, find the minimum value of n. (Original)

Solution. Consider special cases. From the cases where $a = 1, 2, 3, 4$, we find that the minimum value is $n = 1, 2, 2, 3$. Thus, we find that a set $X = \{a_1, a_2, a_3, \ldots, a_n\}$ with the minimum value n while satisfying the requirements has the following property: for any $i = 1, 2, \ldots, n$, $a_i \leqslant 2^{i-1}$.

Actually, suppose that there exists an $i (1 \leqslant i \leqslant n)$ such that $a_i \geqslant 2^{i-1} + 1$, and that i is minimal, then $a_1 \leqslant 2^0$, $a_2 \leqslant 2^1$, $a_3 \leqslant 2^2$, $\ldots$, $a_{i-1} \leqslant 2^{i-2}$, and $a_i \geqslant 2^{i-1} + 1$. Thus for any subset A of X which does not contain any elements in $\{a_i, a_{i+1}, a_{i+2}, \ldots, a_n\}$, we have $S(A) \leqslant a_1 + a_2 + \cdots + a_{i-1} \leqslant 2^0 + 2^1 + 2^2 + \cdots + 2^{i-2} = 2^{i-1} - 1$. For any subset A of X which contains at least one elements of $\{a_i, a_{i+1}, a_{i+2}, \ldots, a_n\}$, $S(A) \geqslant a_i \geqslant 2^{i-1} + 1$. Thus, there exists no subset A of X such that $S(A) = 2^{i-1}$, thus $a \leqslant 2^{i-1} - 1$.

Since the sums of subsets of X cover all of $1, 2, \ldots, a$, and for any subset A of X containing at least one element of $\{a_i, a_{i+1}, a_{i+2}, \ldots, a_n\}$ we have $S(A) > a$, we know that the sums of subsets of $X \backslash \{a_i, a_{i+1}, a_{i+2}, \ldots, a_n\}$ cover $1, 2, \ldots, a$, which contradicts the minimality of n.

For any given a, suppose $2^r \leqslant a < 2^{r+1}$. If $n \leqslant r$, since $a_i \leqslant 2^{i-1}$ $(i = 1, 2, \ldots, n)$, for any subset A of X, we have

$$S(A) \leqslant S(X) = a_1 + a_2 + \cdots + a_n \leqslant 2^0 + 2^1 + 2^2 + \cdots + 2^{n-1}$$
$$= 2^n - 1 \leqslant 2^r - 1 < 2^r \leqslant a,$$

thus there is no subset A of X such that $S(A) = a$, which is a contradiction. So $n \geqslant r + 1$.

When $n = r + 1$, let $a_i = 2^{i-1}$ $(i = 1, 2, \ldots, r)$, $a_{r+1} = a + 1 - 2^r$, we prove that $X = \{a_1, a_2, a_3, \ldots, a_{r+1}\}$ satisfies the requirements.

Actually, using binary representation, we know that the sums of subsets of $\{a_1, a_2, a_3, \ldots, a_{r1}\}$ cover all of $1, 2, \ldots, 2^r - 1$. The subsets of X including a_{r+1} will cover $a_{r+1}, a_{r+1} + 1, a_{r+1} + 2, \ldots, a_{r+1} + 2^r - 1 = a$.

Since $2^r \leqslant a < 2^{r+1}$, we know that $a_{r+1} = a + 1 - 2^r < 2^{r+1} + 1 - 2^r = 2^r + 1$, so $a_{r+1} \leqslant 2^r$.

Thus the sums of subsets of X cover $1, 2, \ldots, a$.

In conclusion, the minimum value of n is $r + 1$, where $r = [\log_2 a]$.

Exercise 13

(1) There are five multiple choice questions in an exam, each question having four different choices, and only one correct answer. In 2000 answer sheets it is found that there is a number n such that for any n answer sheets, there always exist four answer sheets so that for each two of the four, they contain the same answer for at most three questions. Find the minimum value of n. (2000 China Mathematical Olympiad)

(2) Assume that a_1, a_2, ... , a_k is a sequence of positive integers not exceeding n. In any term of the sequence, the two terms beside it are different, and there does not exist four indices $p < q < r < s$, such that $a_p = a_r \neq a_q = a_s$. Find the maximum value of k.

(3) Assume that there are 2^n sequences consisting of $\{0, 1\}$ with finite length, and that none of the sequences is the prefix of another sequence. Find the minimum of S, the sum of the lengths of all sequences.

(4) r stones are placed on an $m \times n (m > 1$, $n > 1)$ chessboard. Each cell of the chessboard has at most one stone. Suppose that the r stones have a property P: each row and each column of the chessboard has at least one stone. Also, if any stone is removed from the board, then P does not hold. Find the maximum value of r.

(5) Let $F = \{A_1, A_2, \ldots, A_k\}$ be a set of subsets of the set $X = \{1, 2, \ldots, n\}$ which satisfies: (1) $| A_i | = 3$; (2) $| A_i \cap A_j | \leqslant 1$. Denote the maximum value of $| F |$ by $f(n)$. Prove that: $\dfrac{n^2 - 4n}{6} \leqslant f(n) \leqslant \dfrac{n^2 - n}{6}$. (The 6th Balkan Mathematical Olympiad)

(6) In an $m \times n (m > 1$, $n > 1)$ chessboard C, each cell is filled with a number, such that for any positive integer p, q and any $p \times q$ rectangle, the sums of the cells in the two diagonals of the rectangle equal each other. If after some r cells are filled, the numbers in the remaining cells can be uniquely determined, find the minimum value of r. (The 5th USSR Mathematical Olympiad)

Solutions to Exercises

Exercise 1

(1) Suppose $a < b < c$, then the minimum and maximum among the seven numbers considered will be $a + b - c$ and $a + b + c$ respectively, thus $d = (a + b + c) - (a + b - c) = 2c$. Note also $a + b - c > 0$, so $c < a + b < a + c < b + c$; but one of $\{a + b, a + c, b + c\}$ equals 800, so $c < 800$. Since $799 = 17 \times 47$ and 798 are both composite, we then have $c \leqslant 797$ and hence $d = 2c \leqslant 1594$.

Next, let $c = 797$ and $a + b = 800$. Noticing that $b < c = 797$, and that the smallest prime solution to $a + b = 800$ is $(a, b) = (13, 787)$ (since $795, 793 = 13 \times 61$ and $799 = 3 \times 263$ are all composite), we may choose $(a, b, c) = (13, 787, 797)$, so that $a + b - c = 3$, $a - b + c = 23$, $-a + b + c = 1571$, $a + b + c = 1597$ are all prime. Therefore, the maximum possible value of d is 1594.

(2) If $n = 1$, then $(a_2 - a_1)^2 = 1$, so $a_2 - a_1 = \pm 1$. In this case the desired maximum is 1. If $n \geqslant 2$, let $x_1 = a_1$ and $x_{i+1} = a_{i+1} - a_i$, for $i = 1, 2, \ldots, 2n - 1$, then $\sum_{i=2}^{2n} x_i^2 = 1$, and $a_k = x_1 + x_2 + \cdots + x_k$ ($k = 1, 2, \ldots, 2n$), so by the Cauchy inequality, we obtain

$$(a_{n+1} + a_{n+2} + \cdots + a_{2n}) - (a_1 + a_2 + \cdots + a_n)$$
$$= n(x_1 + x_2 + \cdots + x_n) + n x_{n+1} + (n - 1) x_{n+2} + \cdots + x_{2n}$$
$$\quad - [n x_1 + (n - 1) x_2 + \cdots + x_n]$$
$$= x_2 + 2x_3 + \cdots + (n - 1) x_n + n x_{n+1} + (n - 1) x_{n+2} + \cdots + x_{2n}$$
$$\leqslant \sqrt{1^2 + 2^2 + \cdots + (n-1)^2 + n^2 + (n-1)^2 + \cdots + 1^2} \sqrt{x_2^2 + x_3^2 + \cdots + x_{2n}^2}$$
$$= \sqrt{n^2 + 2 \times \frac{(n-1)n(2(n-1) + 1)}{6}} = \sqrt{\frac{n(2n^2 + 1)}{3}}.$$

Equality holds when $a_k = \dfrac{\sqrt{3}\,k(k-1)}{2\sqrt{n(2n^2+1)}}$ $(k = 1, 2, \ldots, n + 1)$ and

$$a_{n+k} = \frac{\sqrt{3}\,(2n^2 - (n-k)(n-k+1))}{2\sqrt{n(2n^2+1)}}(k = 1, 2, \ldots, n-1),$$

thus the maximum of $(a_{n+1} + a_{n+2} + \cdots + a_{2n}) - (a_1 + a_2 + \cdots + a_n)$ is $\sqrt{\dfrac{n(2n^2+1)}{3}}$.

(3) The key to solving the problem is to remove the absolute value signs. Note that $|a_i - i|$ equals $a_i - i$ or $i - a_i$, we know that after removing the absolute value signs, there will always be n positive terms and n negative terms. Therefore, if n is even then

$$S_n \leqslant n + n + (n-1) + (n-1) + \cdots + \left(\frac{n}{2} + 1\right) + \left(\frac{n}{2} + 1\right) - \frac{n}{2}$$

$$- \frac{n}{2} - \cdots - 1 - 1 = \frac{n^2}{2}.$$

If n is odd, we have similar results. In any case we have $S_n \leqslant \left[\dfrac{n^2}{2}\right]$. Equality holds when $(a_1, a_2, \ldots, a_n) = (n, n-1, n-2, \ldots, 2, 1$; thus the maximum of S_n is $\left[\dfrac{n^2}{2}\right]$.

(4) For each $k\,(1 \leqslant k \leqslant 1990)$, we have

$$|y_k - y_{k+1}| = \left|\frac{x_1 + x_2 + \cdots + x_k}{k} - \frac{x_1 + x_2 + \cdots + x_{k+1}}{k+1}\right|$$

$$= \left|\frac{x_1 + x_2 + \cdots + x_k - kx_{k+1}}{k(k+1)}\right|$$

$$= \frac{1}{k(k+1)} |(x_1 - x_2) + 2(x_2 - x_3) + 3(x_3 - x_4) + \cdots$$
$$+ k(x_k - x_{k+1})|$$

$$\leqslant \frac{1}{k(k+1)} \mid (x_1 - x_2) \mid + \mid 2(x_2 - x_3) \mid + \mid 3(x_3 - x_4) \mid + \cdots$$

$$+ \mid k(x_k - x_{k+1}) \mid$$

$$= \frac{1}{k(k+1)} \sum_{i=1}^{k} i \mid x_i - x_{i+1} \mid,$$

so

$$\sum_{k=1}^{1990} \mid y_k - y_{k+1} \mid \leqslant \sum_{k=1}^{1990} \left[\frac{1}{k(k+1)} \sum_{i=1}^{k} i \mid x_i - x_{i+1} \mid \right]$$

$$= \sum_{k=1}^{1990} \sum_{i \leqslant k} \left[\frac{1}{k(k+1)} \cdot i \mid x_i - x_{i+1} \mid \right]$$

$$= \sum_{i=1}^{1990} \sum_{k \geqslant i} \left[\frac{1}{k(k+1)} \cdot i \mid x_i - x_{i+1} \mid \right]$$

$$= \sum_{i=1}^{1990} (i \mid x_i - x_{i+1} \mid) \sum_{k=i}^{1990} \frac{1}{k(k+1)}$$

$$= \sum_{i=1}^{1990} (i \mid x_i - x_{i+1} \mid) \sum_{k=i}^{1990} \left(\frac{1}{k} - \frac{1}{k+1} \right)$$

$$= \sum_{i=1}^{1990} (i \mid x_i - x_{i+1} \mid) \left(\frac{1}{i} - \frac{1}{1991} \right)$$

$$= \sum_{i=1}^{1990} (\mid x_i - x_{i+1} \mid) \left(1 - \frac{i}{1991} \right)$$

$$\leqslant \sum_{i=1}^{1990} (\mid x_i - x_{i+1} \mid) \left(1 - \frac{1}{1991} \right)$$

$$= 1991 \left(1 - \frac{1}{1991} \right) = 1990.$$

Equality holds when $x_1 = 1991$, and $x_2 = x_3 = \cdots = x_{1991} = 0$. Therefore $F_{\max} = 1990$.

(5) Note that $F_2 = \mid 1 - 2 \mid = 1 \leqslant 2$, $F_3 = \mid \mid 1 - 2 \mid - 3 \mid = 2 \leqslant 3$, $F_4 = \mid \mid \mid 1 - 2 \mid - 3 \mid - 4 \mid = 4 \leqslant 4$. In general, we will conjecture that $F_n \leqslant n$. Notice that for x, $y > 0$, we have $\mid x - y \mid \leqslant \max\{x, y\}$, thus $\mid x_1 - x_2 \mid \leqslant \max\{x_1, x_2\}$, and

$$\mid \mid x_1 - x_2 \mid - x_3 \mid \leqslant \max\{\max\{x_1, x_2\}, x_3\} = \max\{x_1, x_2, x_3\}.$$

Assume

$$|\,|\,x_1 - x_2\,| - x_3 - \cdots - x_k\,| \leqslant \max\{x_1,\, x_2,\, x_3,\, \ldots,\, x_k\},$$

Then

$$|\,|\,x_1 - x_2\,| - x_3 - \cdots - x_{k+1}\,| \leqslant \max\{\max\{x_1,\, x_2,\, x_3,\, \ldots,\, x_k\},\, x_{k+1}\}$$
$$= \max\{x_1,\, x_2,\, x_3,\, \ldots,\, x_k,\, x_{k+1}\}.$$

By induction we have $F = |\,|\,x_1 - x_2\,| - x_3 - \cdots - x_{1990}\,| \leqslant \max\{x_1,\, x_2,\, x_3,\, \ldots,\, x_{1990}\} = 1990$.

However, equality may not hold, as suggested by the initial values; in fact, equality does not hold for $k = 1990$. To see this, note that removing absolute value signs and changing signs of the terms does not affect the parity of F, thus

$$F \equiv x_1 + x_2 + x_3 + \cdots + x_{1990} = 1 + 2 + \cdots + 1990 \equiv 1 (\mathrm{mod}\, 2).$$

This implies $F \leqslant 1989$.

Below we will construct $(x_1,\, x_2,\, x_3,\, \ldots,\, x_{1990})$ so that $F = 1989$. The strategy is to make as many terms in the sum cancel as possible. We may choose $x_{1989} = 1990$, $x_{1990} - 1$, so that we need the other 1988 numbers to cancel each other; recall that for four consecutive integers $n + 1$, $n + 2$, $n + 3$ and $n + 4$, we have

$$|\,|\,|\,n + 3 - (n + 1)\,| - (n + 4)\,| - (n + 2)\,| = 0;$$

thus we can divide $(x_1,\, x_2,\, x_3,\, \ldots,\, x_{1988})$ into groups of four, and let

$$(x_{4k+1},\, x_{4k+2},\, x_{4k+3},\, x_{4k+4}) = (4k + 2,\, 4k + 4,\, 4k + 5,\, 4k + 3),$$

then

$$|\,|\,x_{4k+1} - x_{4k+2}\,| - x_{4k+3} - x_{4k+4}\,|$$
$$= |\,|\,|\,(4k + 2) - (4k + 4)\,| - (4k + 5)\,| - (4k + 3)\,| = 0,$$

and hence $F = |\,1990 - 1\,| = 1989$.

(6) Suppose $a_1 \leqslant a_2 \leqslant \cdots \leqslant a_n$. Since the function under consideration is continuous in a closed domain, it attains its maximum and minimum. By rearrangement inequality, we obtain

$$F = \sum_{i=1}^{n} \frac{a_i}{b_i} \geqslant \sum_{i=1}^{n} \frac{a_i}{a_i} = \sum_{i=1}^{n} 1 = n;$$

equality holds when $a_i = b_i \, (1 \leqslant i \leqslant n)$. Thus $F_{\min} = n$.

Again by rearrangement inequality, we obtain

$$F = \sum_{i=1}^{n} \frac{a_i}{b_i} \leqslant \sum_{i=1}^{n} \frac{a_i}{a_{n+1-i}} = F'.$$

We next find the maximum of $F' = \sum_{i=1}^{n} \frac{a_i}{a_{n+1-i}}$. By symmetry,

$$2F' = \sum_{i=1}^{n} \left(\frac{a_i}{a_{n+1-i}} + \frac{a_{n+1-i}}{a_i} \right).$$

Because $p \leqslant a_i \leqslant q$, $\dfrac{p}{q} \leqslant \dfrac{a_i}{a_{n+1-i}} \leqslant \dfrac{q}{p}$, and $f(x) = x + \dfrac{1}{x}$ is decreasing in $(0, 1]$ and increasing in $[1, \infty)$, we know that $2F'$ attains its maximum only when $\dfrac{a_i}{a_{n+1-i}} \in \left\{ \dfrac{p}{q}, \dfrac{q}{p} \right\}$. When n is even, each of $\dfrac{a_i}{a_{n+1-i}}$ can attain the value $\dfrac{p}{q}$ or $\dfrac{q}{p}$, so

$$F_{\max} = \frac{n}{2} \cdot \left(\frac{p}{q} + \frac{q}{p} \right) = n + \left[\frac{n}{2} \right] \left(\sqrt{\frac{p}{q}} - \sqrt{\frac{q}{p}} \right)^2.$$

When n is odd, we always have $\dfrac{a_{\left[\frac{n}{2} \right] + 1}}{a_{n+1-\left(\left[\frac{n}{2} \right] + 1 \right)}} = \dfrac{a_{\left[\frac{n}{2} \right] + 1}}{a_{n - \left[\frac{n}{2} \right]}} = 1$, and the others of $\dfrac{a_i}{a_{n+1-i}}$ can attain $\dfrac{p}{q}$ or $\dfrac{q}{p}$. In this case,

$$F_{\max} = \frac{n-1}{2} \cdot \left(\frac{p}{q} + \frac{q}{p} \right) + 1 = n + \left[\frac{n}{2} \right] \left(\sqrt{\frac{p}{q}} - \sqrt{\frac{q}{p}} \right)^2.$$

In summary the minimum value of F is n, the maximum value of F is $n + \left[\dfrac{n}{2} \right] \left(\sqrt{\dfrac{p}{q}} - \sqrt{\dfrac{q}{p}} \right)^2$.

Exercise 2

(1) Since $x + y + z = a$, we have $\dfrac{x}{a} + \dfrac{y}{a} + \dfrac{z}{a} = 1$. Let $x = au$, $y = $

av, $z = aw$, then $0 \leqslant u$, v, $w \leqslant 1$, $u + v + w = 1$, $F = 2x^2 + y + 3z^2 = 2a^2u^2 + av + 3a^2w^2$. Since $a^2 \geqslant a \geqslant 1$ for $a \geqslant 1$, we have

$$F = 2a^2u^2 + av + 3a^2w^2 \leqslant 2a^2u^2 + a^2v + 3a^2w^2$$
$$\leqslant 3a^2u^2 + 3a^2v + 3a^2w^2 \leqslant 3a^2u + 3a^2v + 3a^2w = 3a^2.$$

Equality holds when $u = v = 0$, $w = 1$, i.e. $x = y = 0$, and $z = a$. Therefore the maximum value of $2x^2 + y + 3z^2$ is $3a^2$.

(2) Assume the required number is $100x + 10y + z$, consider

$$F = \frac{100x + 10y + z}{x + y + z} = 1 + \frac{99x + 9y}{x + y + z}.$$

On the right-hand side, z only appears in the denominator; thus F is decreasing in z. Fixing x, y, since $z \leqslant 9$, we know

$$F \geqslant 1 + \frac{99x + 9y}{x + y + 9} = 10 + \frac{90x - 81}{x + y + 9}.$$

On the right-hand side, y only appears in the denominator, thus this expression is decreasing in y. Fixing x, since $y \leqslant 9$, we know

$$F \geqslant 10 + \frac{90x - 81}{x + 9 + 9} = 100 - \frac{1701}{18 + x} \geqslant 100 - \frac{1701}{19} = \frac{199}{19}.$$

Equality holds when $x = 1$, $y = z = 9$, so the required number is 199.

(3) First, consider $g(x) = \dfrac{a}{x + b} + \dfrac{x}{(x + b)^2}$ with $a \geqslant 0$ and $b \geqslant 1$. By elementary arguments, we know that the maximum of g is $\dfrac{(1 + a)^2}{4b}$, attained at $x = \dfrac{b(1 - a)}{1 + a}$. Now, with fixed $(x_2, x_3, \ldots, x_n)$, then H is a function of form $g(x_1) + C_1$, where $a_1 = 0$, $b_1 = 1 + x_2 + \cdots + x_n$. By the above arguments, the maximum value of $g(x_1) + C_1$ is

$$\frac{(1 + a_1)^2}{4} \frac{1}{1 + x_2 + x_3 + \cdots + x_n} + \frac{x_2}{(1 + x_2 + x_3 + \cdots + x_n)^2}$$

$$+\cdots+\frac{x_n}{(1+x_n)^2}=H_2,$$

attained at $x_1=\dfrac{(1+x_2+x_3+\cdots+x_n)(1-a_1)}{1+a_1}$. Now fix $(x_3,\ x_4,\ \ldots,$

$x_n)$, then H_2 is a function of form $g(x_2)+C_2$, where $a_2=\dfrac{(1+a_1)^2}{4}$,

$b_2=1+x_3+\cdots+x_n$.

Again, the maximum value of $g(x_2)+C_2$ is

$$\frac{(1+a_2)^2}{4}\frac{1}{1+x_3+x_4+\cdots+x_n}+\frac{x_3}{(1+x_3+x_4+\cdots+x_n)^2}$$

$$+\cdots+\frac{x_n}{(1+x_n)^2}=H_3,$$

attained at $x_2=\dfrac{(1+x_3+x_4+\cdots+x_n)(1-a_2)}{1+a_2}$. In this way, we

finally obtain

$$\frac{(1+a_{n-1})^2}{4}\frac{1}{1+x_n}+\frac{x_n}{(1+x_n)^2}=H_n.$$

By the above arguments with $b=1$, we know that the maximum value

of H_n is $\dfrac{(1+a_n)^2}{4}$, attained at $x_n=\dfrac{1-a_n}{1+a_n}$, where $a_n=\dfrac{(1+a_{n-1})^2}{4}$. To

summarize, let the maximum of H be a_n, then we have $a_1=0$ and $a_k=$

$\dfrac{(1+a_{k-1})^2}{4}$. This maximum is attained when $x_n=\dfrac{1-a_n}{1+a_n}$, $x_{n-1}=$

$\dfrac{(1+x_n)(1-a_{n-1})}{1+a_{n-1}}$, $\ldots$, $x_1=\dfrac{(1+x_2+x_3+\cdots+x_n)(1-a_1)}{1+a_1}$. It is

easy to show that $a_n\geqslant a_{n-1}$, and $0\leqslant a_n\leqslant1$; therefore $x_1,\ x_2,\ \ldots,\ x_n$

are all nonnegative. Since $\{a_n\}$ is monotonic and bounded, it must

have a limit a. We then have $a=\dfrac{(1+a)^2}{4}$, thus $a=1$.

(4) Let the maximum of $\dfrac{b}{a}+\dfrac{d}{c}$ be $f(n)$. We may assume $a\leqslant c$.

If $a\geqslant n+1$, then

$$\frac{b}{a} + \frac{d}{c} \leqslant \frac{b}{a} + \frac{d}{a} = \frac{b+d}{a} \leqslant \frac{n}{n+1}.$$

If $a \leqslant n$, then we fix a, and let $x = a(n - a + 1) + 1$. When $c \leqslant x$, from $\frac{b}{a} + \frac{d}{c} < 1$ we know that $bc + ad < ac$, thus $bc + ad \leqslant ac - 1$, so

$$\frac{b}{a} + \frac{d}{c} = \frac{bc + ad}{ac} \leqslant \frac{ac - 1}{ac} = 1 - \frac{1}{ac} \leqslant 1 - \frac{1}{ax}.$$

When $c > x$, from $\frac{b}{a} + \frac{d}{c} < 1$ we have that $\frac{b}{a} < 1$, hence $a \geqslant b + 1$, therefore

$$\frac{b}{a} + \frac{d}{c} - \left(\frac{a-1}{a} + \frac{b+d-a+1}{c} \right) = (b+1-a)\left(\frac{1}{a} - \frac{1}{c} \right) \leqslant 0.$$

It follows that

$$\frac{b}{a} + \frac{d}{c} \leqslant \frac{a-1}{a} + \frac{b+d-a+1}{c} \leqslant \frac{a-1}{a} + \frac{n-a+1}{c}$$

$$\leqslant \frac{a-1}{a} + \frac{n-a+1}{x} = 1 - \frac{1}{ax}.$$

In any case we have $\frac{b}{a} + \frac{d}{c} \leqslant 1 - \frac{1}{ax}$, so we only need to find the maximum of the function

$$g(a) = 1 - \frac{1}{ax} = 1 - \frac{1}{a[a(n-a+1)+1]} \quad (2 \leqslant a \leqslant n, \ a \in \mathbf{N}^*),$$

or equivalently, finding the maximum of the function

$$h(a) = a[a(n-a+1)+1] \quad (2 \leqslant a \leqslant n, \ a \in \mathbf{N}^*).$$

Since $h'(a) = -3a^2 + (2n+2)a + 1$, the positive root of $h'(a) = 0$ is $a_0 = \frac{n+1+\sqrt{(n+1)^2+3}}{3}$, and $\frac{2n+2}{3} < a_0 < \frac{2n+3}{3} = \frac{2n}{3} + 1$, we find that the maximum of $h(a)$ is attained at $a = \left[\frac{2n}{3} \right] + 1$, so the desired maximum is $1 - \frac{1}{a[a(n-a+1)+1]}$, where $a = \left[\frac{2n}{3} \right] + 1$.

Exercise 3

(1) When 1989 is partitioned into $199 + 199 + \cdots + 199 + 198$, the corresponding product will be $199^9 \times 198$. We will prove that this is the maximum. First the maximum must exist, because the number of partitions is finite; assume that $(x_1, x_2, \ldots, x_{10})$ is a partition where $x_1 x_2 \cdots x_{10}$ is maximized. If one of $(x_1, x_2, \ldots, x_{10})$ is less than 198, assume it is x_1, then there must also be a number larger than 198, say x_{10}. Let $x_1' = x_1 + (x_{10} - 198)$, $x_{10}' = 198$, we obtain a new partition $(x_1', x_2, \ldots, x_9, x_{10}')$, with the corresponding product $x_1' x_2 \cdots x_9 x_{10}'$. Since

$$
\begin{aligned}
x_1' x_{10}' - x_1 x_{10} &= [x_1 + (x_{10} - 198)]198 - x_1 x_{10} \\
&= 198 x_1 + 198 x_{10} - 198^2 - x_1 x_{10} \\
&= (198 - x_1)(x_{10} - 198) > 0,
\end{aligned}
$$

we have $x_1' x_2 \cdots x_9 x_{10}' > x_1 x_2 \cdots x_{10}$, which is a contradiction. Thus each of $(x_1, x_2, \ldots, x_{10})$ is not less than 198. Similarly, if one of $(x_1, x_2, \ldots, x_{10})$ is larger than 199, assume it is x_1, then there must also be a number less than 199, say x_{10}. Since x_{10} is also not less than 198, we know $x_{10} = 198$. Let $x_1' = x_1 - 1$, $x_{10}' = x_{10} + 1$, we obtain a new partition $(x_1', x_2, \ldots, x_9, x_{10}')$ with the corresponding product $x_1' x_2 \cdots x_9 x_{10}'$. But

$$
\begin{aligned}
x_1' x_{10}' - x_1 x_{10} &= (x_1 - 1)(x_{10} + 1) - x_1 x_{10} \\
&= x_1 x_{10} + x_1 - x_{10} - 1 - x_1 x_{10} = x_1 - x_{10} - 1 \\
&= x_1 - 198 - 1 = x_1 - 199 > 0.
\end{aligned}
$$

Thus $x_1' x_2 \cdots x_9 x_{10}' > x_1 x_2 \cdots x_{10}$, which is a contradiction. Therefore each of $(x_1, x_2, \ldots, x_{10})$ must be either 199 or 198, so we obtain the partition $199 + 199 + \cdots + 199 + 198$, proving that this product $199^9 \times 198$ is indeed maximal.

(2) First, the maximum must exist. By assumption, we have $a_1 \leqslant$

$a_2 \leqslant \cdots \leqslant a_{19} = 85$. We guess that at maximum, each a_i is as large as possible and all a_i are equal, and all b_j are also equal. In fact, if there exists an i such that $a_i < a_{i+1} (1 \leqslant i \leqslant 18)$, then let $a_i' = a_i + 1$, $a_j' = a_j$ for all $j \neq i$, and denote the corresponding b_j by b_j'. Since $a_{i+1} > a_i$, we know $a_{i+1} \geqslant a_i + 1$, but $a_i < a_i + 1$, thus $b_{a_i+1} = i + 1$, $b_{a_i+1}' = i = b_{a_i+1} - 1$, $b_j' = b_j$ (when $j \neq a_i + 1$). Therefore, b_{a_i+1} is decreased by 1, and any other b_i remains unchanged; thus the value of S is not changed. In summary, we have $S \leqslant 19 \times 85 + 1 \times 85 = 1700$, and equality holds when $a_i = 85$, $b_j = 1(1 \leqslant i \leqslant 19, 1 \leqslant j \leqslant 85)$.

(3) The problem is equivalent to dividing 155 birds into a number of groups, so that the number of mutually visible bird pairs is minimized; note that the number of groups is not fixed. By considering special cases, we know that in order to minimize the number of visible bird pairs, any two adjacent bird positions should not be too close; i. e., any two positions should not be mutually visible.

To be precise, if (P_i, P_j) is a visible bird pair, we call the position of (P_i, P_j) a visible position pair. Suppose $(P_i, P_j,)$ is such a pair, let k be the number of the birds that P_i cannot see and P_j can see, t be the number of the birds that P_j cannot see and P_i can see. Suppose $k \geqslant t$, then we can move all birds at the position of P_j to the position of P_i. For any bird pair (P, Q), if it does not contain a bird that is moved, its "visibility" will remain unchanged; otherwise if P is moved, then the pair changes from "*visible*" to "*invisible*" for exactly k choices of Q, and changes from "*invisible*" to "*visible*" for exactly t choices of Q. Therefore, the number of visible bird pairs does not increase. Since each operation decreases the number of different bird positions by 1, we eventually arrive at the situation where any position pair is not visible. This means that there are at most 35 bird positions, so the problem is reduced to finding the maximum value of

$$S = \sum_{i=1}^{35} C_{x_i}^2 = \frac{1}{2} \sum_{i=1}^{35} x_i (x_i - 1),$$

where $x_1 + x_2 + \cdots + x_{35} = 155$, $x_i \geqslant 0$.

We cannot have all x_i equal, since 35 does not divide 155; but we do have $|x_i - x_j| \leqslant 1$ for each i, j when the maximum is attained. In fact, if $x_i - x_j \geqslant 2$, we may assume that $x_2 - x_1 \geqslant 2$, then let $x_1' = x_1 + 1$, $x_2' = x_2 - 1$. In this case,

$$x_1(x_1 - 1) + x_2(x_2 - 1) - [x_1'(x_1' - 1) + x_2'(x_2' - 1)]$$
$$= x_1(x_1 - 1) + x_2(x_2 - 1) - (x_1 + 1)x_1 - (x_2 - 1)(x_2 - 2)$$
$$= -x_1 + (x_2 - 1) \geqslant 1.$$

So S is made strictly smaller. Notice that $155 = 4 \times 35 + 15$, we know that the point of minimum is $(x_1, x_2, \ldots, x_{35}) = (5, 5, \ldots, 5, 4, 4, \ldots, 4)$, and the minimum of S is $20 \, C_4^2 + 15 C_5^2 = 270$.

(4) Heuristically, the point of extremum $(x_1, x_2, \ldots, x_n)$ should be homogeneous, i. e., $x_1 = x_2 = \cdots = x_n = x$ (to be determined). In this case,

$$d = \sum_{i=1}^{n} (P_i - x_i)^2 = \sum_{i=1}^{n} (P_i - x)^2 = nx^2 - 2 \sum_{i=1}^{n} (P_i)x + \sum_{i=1}^{n} P_i^2.$$

The minimum of this quadratic function is attained at

$$x = \sum_{i=1}^{n} \frac{p_i}{n} = p.$$

So we conjecture that the minimum of d is attained at $x_1 = x_2 = \cdots = x_n = p$.

We only need to prove that for real numbers $P_1 \leqslant P_2 \leqslant P_3 \leqslant \cdots \leqslant P_n$, and $x_1 \geqslant x_2 \geqslant \cdots \geqslant x_n$, we have $\sum_{i=1}^{n} (P_i - x_i)^2 \geqslant \sum_{i=1}^{n} (P_i - p)^2$, where $p = \sum_{i=1}^{n} \frac{P_i}{n}$. Denote by H the difference between the left- and right-hand sides, then

$$H = \sum_{i=1}^{n} x_i^2 - 2 \sum_{i=1}^{n} P_i x_i + 2p \sum_{i=1}^{n} P_i - np^2$$
$$= \sum_{i=1}^{n} x_i^2 - 2 \sum_{i=1}^{n} P_i x_i + np^2 \geqslant \sum_{i=1}^{n} x_i^2 - 2 \cdot \frac{1}{n} \sum_{i=1}^{n} P_i \sum_{i=1}^{n} x_i + np^2$$

$$= \sum_{i=1}^{n} x_i^2 - 2p \sum_{i=1}^{n} x_i + np^2 = \sum_{i=1}^{n} (x_i - p)^2 \geqslant 0,$$

by the Chebyshev inequality.

(5) (i) Since the number of partitions is finite, there must exist a partition so that the number of triangles is maximized. Notice that $1994 = 83 \times 24 + 2 = 81 \times 24 + 2 \times 25$, thus the 1994 points be partitioned into 83 groups, with 81 groups having 24 points and 2 groups having 25 points. We will prove that this minimizes the number of triangles. In fact, if in a minimizing partition, two groups have i and j points respectively, with $i \geqslant j + 2$, then we may move a point from the group with i points to the group with j points, so that the increment of S is

$$-S + S' = -C_i^3 - C_j^3 + (C_{i-1}^3 + C_{j+1}^3) = -C_{i-1}^2 + C_j^2 < 0,$$

which is a contradiction. (ii) From above, we know that G' consists of a number of independent connected graphs, thus we only need to consider the coloring of two graphs G_1 and G_2 with $| G_1 | = 25$ and $| G_2 | = 24$; actually we only need to consider G_1. We divide the 25 points into five groups, each including five points. For each group, we color the five points using the above-described method; this five-point group is then viewed as a "*big point*", so that there are five "*big points*". Edges connecting these five points are then colored using the remaining two colors using the method above, so that the graph is colored with four colors.

(6) **Solution 1.** Suppose 14 peoples are A_1, A_2, ..., A_{14}, and the numbers of wins for each them are w_1, w_2, ..., w_{14} respectively, then $\sum_{i=1}^{14} w_i = C_{14}^2 = 91$. For any three people not forming a triangle, there must be the one who defeats the two others thus there are $C_{w_i}^2$ triples containing A_i and not being "*triangles*", therefore the total number of nontriangle triples is $\sum_{i=1}^{14} C_{w_i}^2$ (we define $C_0^2 = C_1^2 = 0$),

thus the total number of triangles is $S = C_{14}^3 - \sum_{i=1}^{14} C_{w_i}^2$.

Next we will find the minimum value of $S' = \sum_{i=1}^{14} C_{w_i}^2$. First the minimum must exist by finiteness; next, when the minimum is achieved, then for any $1 \leqslant i < j \leqslant 14$, we must have $|w_i - w_j| \leqslant 1$. In fact, if for some $1 \leqslant i < j \leqslant 14$ we have $w_i - w_j \geqslant 2$, then let $y_i = w_i - 1$, $y_j = w_j + 1$, $y_k = w_k (k \neq i, j)$, then

$$\sum_{i=1}^{14} C_{w_i}^2 - \sum_{i=1}^{14} C_{y_i}^2 = C_{w_i}^2 + C_{w_j}^2 - (C_{w_i-1}^2 + C_{w_j+1}^2) = w_i - w_j - 1 > 0,$$

which is a contradiction. Notice that $91 = 14 \times 6 + 7$, so at the point of minimum we have

$$\{w_1, w_2, \ldots, w_{14}\} = \{6, 6, 6, 6, 6, 6, 6, 7, 7, 7, 7, 7, 7, 7\},$$

and the minimum value is $7C_6^2 + 7C_7^2 = 252$, therefore $S = C_{14}^3 - \sum_{i=1}^{14} C_{w_i}^2 \leqslant C_{14}^3 - 252 = 112$.

Solution 2. As with Solution 1, we know that the total number of nontriangle triples is $\sum_{i=1}^{14} C_{w_i}^2$ (which defines $C_0^2 = C_1^2 = 0$). Because the number of losses of A_i is l_i, similarly we have that the total number of nontriangle triples also equals $\sum_{i=1}^{14} C_{l_i}^2$. Therefore, the total number of nontriangle triples is $\frac{1}{2} \sum_{i=1}^{14} (C_{l_i}^2 + C_{w_i}^2)$. Since $w_i + l_i = 13$, we have

$$w_i^2 + l_i^2 = \frac{1}{2}[13^2 + (w_i - l_i)^2] \geqslant 85,$$

so $C_{w_i}^2 + C_{l_i}^2 = \frac{w_i^2 + l_i^2}{2} - \frac{13}{2} \geqslant 36$, and hence $\frac{1}{2} \sum_{i=1}^{14} (C_{l_i}^2 + C_{w_i}^2) \geqslant \frac{1}{2} \sum_{i=1}^{14} 36 = 252$. Therefore the number of triangles satisfies $S \leqslant C_{14}^3 - 252 = 112$. Finally, for any $1 \leqslant i < j \leqslant 14$, let A_i defeat A_j if and only if i, j have the same parity, then $w_2 = w_4 = w_6 = w_8 = w_{10} = w_{12} = w_{14} = 7$, $w_1 = w_3 = w_5 = w_7 = w_9 = w_{11} = w_{13} = 6$, and $S = 112$. In summary, the maximum requirements is 112.

Exercise 4

(1) The maximum value of F is: $n^2 + \left[\dfrac{n}{2}\right]\left[\dfrac{n+1}{2}\right]\left(\sqrt{\dfrac{p}{q}} - \sqrt{\dfrac{q}{p}}\right)^2$. Since F is continuous in a closed domain, there must exist a maximum. Fix $a_1, a_2, \ldots, a_{n-1}$, denote

$$a_1 + a_2 + \cdots + a_{n-1} = A, \quad \frac{1}{a_1} + \frac{1}{a_2} + \cdots + \frac{1}{a_{n-1}} = B,$$

where A, B are constants, then

$$F = (A + a_n)\left(B + \frac{1}{a_n}\right) = 1 + AB + Ba_n + \frac{A}{a_n}.$$

Consider at $f(x) = Bx + \dfrac{A}{x}$, it is easy to know that when $x \leqslant \sqrt{\dfrac{A}{B}}$, $f(x)$ is monotonically decreasing, and when $x \geqslant \sqrt{\dfrac{A}{B}}$, $f(x)$ is monotonically increasing. Thus $\sqrt{\dfrac{A}{B}}$ is the minimum point of $f(x)$.

Notice that $p \leqslant a_n \leqslant q$, so $f(a_n) = Ba_n + \dfrac{A}{a_n}$ reaches the maximum only at the end-points: $a_n = p$ or $a_n = q$. By symmetry, F reaches the maximum only when $a_1, a_2, \ldots, a_n \in \{p, q\}$. Therefore, we can assume that there are k numbers among $a_1, a_2, \ldots, a_n$ that equal p and $n - k$ numbers of $a_1, a_2, \ldots, a_n$ that equal q when F reaches the maximum value, thus the maximum value of F is the maximum value of $F(k)$ with the following form:

$$F(k) = [kp + (n-k)q]\left(\frac{k}{p} + \frac{n-k}{q}\right)$$

$$= k^2 + (n-k)^2 + (nk - k^2)\left(\frac{p}{q} + \frac{q}{p}\right)$$

$$= \left[2 - \left(\frac{p}{q} + \frac{q}{p}\right)\right]k^2 + \left[n\left(\frac{p}{q} + \frac{q}{p}\right) - 2n\right]k + n^2.$$

In the quadratic function $F(k)$, the coefficient of the quadratic term is $2 - \left(\dfrac{p}{q} + \dfrac{q}{p}\right) < 0$, and the axis of symmetry is $x = -n/2$, so when n is even,

$$F(k) \leqslant F\left(\frac{n}{2}\right) = \left(\frac{n}{2}\right)^2 + \left(n - \frac{n}{2}\right)^2 + \left[n \cdot \frac{n}{2} - \left(\frac{n}{2}\right)^2\right]\left(\frac{p}{q} + \frac{q}{p}\right)$$

$$= n^2 + \left(\frac{n}{2}\right)^2 \left(\sqrt{\frac{p}{q}} - \sqrt{\frac{q}{p}}\right)^2$$

$$= n^2 + \left[\frac{n}{2}\right]\left[\frac{n+1}{2}\right]\left(\sqrt{\frac{p}{q}} - \sqrt{\frac{q}{p}}\right)^2.$$

When n is odd,

$$F(k) \leqslant F\left(\left[\frac{n}{2}\right]\right) = F\left(\left[\frac{n+1}{2}\right]\right)$$

$$= n^2 + \left[\frac{n}{2}\right]\left[\frac{n+1}{2}\right]\left(\sqrt{\frac{p}{q}} - \sqrt{\frac{q}{p}}\right)^2.$$

In summary, $F_{\max} = n^2 + \left[\dfrac{n}{2}\right]\left[\dfrac{n+1}{2}\right]\left(\sqrt{\dfrac{p}{q}} - \sqrt{\dfrac{q}{p}}\right)^2.$

(2) Since F is continuous in a closed domain, there must exist a minimum. Fix x_2, x_3, ..., x_n, then $F(x_1)$ is a linear function of x_1. Notice that $-1 \leqslant x_1 \leqslant 1$, so $x_1 \in \{-1, 1\}$ when $F(x_1)$ reaches the minimum value. By symmetry, we known $x_i \in \{-1, 1\}(1 \leqslant i \leqslant n)$ when F reaches the minimum value. By this reason, we can assume there are k numbers among x_1, x_2, ..., x_n that equal 1 and $n - k$ numbers among x_1, x_2, ..., x_n that equal -1 when F reaches the minimum value. Therefore the minimum value of F is the minimum value of $F(k)$ with the following form:

$$F(k) = C_k^2 + C_{n-k}^2 - C_k^1 C_{n-k}^1 = 2\left(k - \frac{n}{2}\right)^2 - \frac{n}{2} \geqslant -\frac{n}{2}.$$

Since $F(k)$ is an integer, we know $F(k) \geqslant -\left[\dfrac{n}{2}\right]$, and equality holds if $k = \left[\dfrac{n}{2}\right]$. Therefore, the minimum value of F is $-\left[\dfrac{n}{2}\right]$. We can also solve as follows: when $x_i \in \{-1, 1\}$,

$$F = \frac{1}{2}\left(\sum_{i=1}^{n}x_i\right)^2 - \sum_{i=1}^{n}\frac{1}{2}x_i^2 = \frac{1}{2}\left(\sum_{i=1}^{n}x_i\right)^2 - \sum_{i=1}^{n}\frac{1}{2}$$

$$= \frac{1}{2}\left(\sum_{i=1}^{n}x_i\right)^2 - \frac{n}{2} \geqslant -\frac{n}{2},$$

so $-F \leqslant \dfrac{n}{2}$; but $-F$ is an integer, so $-F \leqslant \left[\dfrac{n}{2}\right]$, i.e., $F \geqslant -\left[\dfrac{n}{2}\right]$.

(3) Since F is continuous in a closed domain, there must exist a maximum and a minimum. Assume the extremum point of F is $(x_1, x_2, \ldots, x_n)$, we will prove that $x_ix_j = 0$ or $x_i = x_j$ for any $i \neq j$. In fact, by symmetry we just need to consider x_1, x_2. Fix x_3, x_4, $\ldots$, x_n, then $x_1 + x_2 = 1 - (x_3 + x_4 + \cdots + x_n) = c$ (constant), and

$$F = (x_1 + x_2)^2 + x_3^2 + \cdots + x_n^2 - 2x_1x_2 + \lambda x_1x_2\cdots x_n$$
$$= (x_1 + x_2)^2 + x_3^2 + \cdots + x_n^2 + (\lambda x_3x_4\cdots x_n - 2)x_1(c - x_1).$$

Define $f(x_1) = x_1(c - x_1)$, since $0 \leqslant x_1 \leqslant c$, by the property of a quadratic function, we know $x_1 \in \left\{0, c, \dfrac{c}{2}\right\}$ when $f(x_1)$ achieves its extremum value.

Notice that at this time we have $x_1 + x_2 = c$, thus we have one of the following conditions: (i) $x_1 = 0$, $x_2 = c$. (ii) $x_1 = c$, $x_2 = 0$. (iii) $x_1 = x_2 = \dfrac{c}{2}$. Therefore, either $x_1x_2 = 0$, or $x_1 = x_2$. From the discussion above, it is clear that if $(x_1, x_2, \ldots, x_n)$ is the extremum point of F, then the nonzero numbers among $x_1, x_2, \ldots, x_n$ are equal. Assume that $x_1 = x_2 = \cdots = x_k \neq 0$, and $x_{k+1} = x_{k+2} = \cdots = x_n = 0$, we consider two cases: (i) If $k = n$, then F achieves the extremum value at $x_1 = x_2 = \cdots = x_n = \dfrac{1}{n}$, and

$$F = \sum_{i=1}^{n}\frac{1}{n^2} + \lambda\prod_{i=1}^{n}\frac{1}{n} = \frac{\lambda + n^{n-1}}{n^n}.$$

(ii) If $k < n$, then F achieves the extremum value at $x_1 = x_2 = \cdots = x_k = \frac{1}{k}$, $x_{k+1} = x_{k+2} = \cdots = x_n = 0$, and $F = \sum_{i=1}^{k} \frac{1}{k^2} = \frac{1}{k}$. Notice that $1 \leqslant k \leqslant n - 1$, thus $\frac{1}{n-1} \leqslant F \leqslant 1$. Therefore the set of the extremum value of F is $\left\{ 1, \frac{1}{n-1}, \frac{\lambda + n^{n-1}}{n^n} \right\}$. In summary, $F_{\min} = \min\left\{ \frac{1}{n-1}, \frac{\lambda + n^{n-1}}{n^n} \right\}$, so

$$F_{\max} = \max\left\{ 1, \frac{\lambda + n^{n-1}}{n^n} \right\}.$$

Exercise 5

(1) If $n = 2$, then $x_1 + x_2 \leqslant \frac{1}{2}$, $(1 - x_1)(1 - x_2) = 1 + x_1 x_2 - (x_1 + x_2) \geqslant 1 - x_1 - x_2 \geqslant \frac{1}{2}$. Equality holds if $x_1 + x_2 = \frac{1}{2}$, $x_1 x_2 = 0$, i.e. $x_1 = \frac{1}{2}$, $x_2 = 0$. If $n = 3$, then $x_1 + x_2 + x_3 \leqslant \frac{1}{2}$. Applying the above transformation we get

$$(1 - x_1)(1 - x_2)(1 - x_3) \geqslant (1 - x_1)[1 - (x_2 + x_3)]$$
$$\geqslant 1 - x_1 - (x_2 + x_3) \geqslant \frac{1}{2},$$

Where we have noticed that $0 \leqslant x_i < 1$, $1 - x_i > 0$. Equality holds if $x_1 + x_2 + x_3 = \frac{1}{2}$, and $x_2 x_3 = x_1(x_2 + x_3) = 0$, i.e. $x_1 = \frac{1}{2}$, $x_2 = x_3 = 0$. We can guess from the above that the extremum point in the general case is: $\left(\frac{1}{2}, 0, 0, \ldots, 0 \right)$. We shall prove this using polishing transformation. First prove the lemma: If $0 \leqslant x, y \leqslant 1$, then $(1 - x)(1 - y) \geqslant 1 - x - y$. This is proved by expanding the left-hand side. This polishing tool can be written as $(x, y) \to (x + y, 0)$. Suppose $n \geqslant 2$, we may assume that $x_1 \geqslant x_2 \geqslant \cdots \geqslant x_n$, then

$$F = (1 - x_1)(1 - x_2)\cdots(1 - x_n)$$
$$\geqslant (1 - x_1)(1 - x_2)\cdots(1 - x_{n-2})(1 - x_{n-1} - x_n)$$
$$\geqslant (1 - x_1)(1 - x_2)\cdots(1 - x_{n-3})(1 - x_{n-2} - x_{n-1} - x_n)$$
$$\geqslant \cdots \geqslant 1 - x_1 - x_2 - \cdots - x_n \geqslant \frac{1}{2}.$$

Equality holds if the variables are $\left(\frac{1}{2}, 0, 0, \ldots, 0\right)$, and F reaches

the minimum of $\frac{1}{2}$.

(2) Using the method of the previous problem, we can obtain the
maximum value of F is $F\left(\frac{1}{2}, \frac{1}{2}, 0, 0, \ldots, 0\right) = \frac{1}{4}$.

(3) If $n = 2$, then $x_1 + x_2 = \pi$, $F = \sin^2 x_1 + \sin^2 x_2 = 2\sin^2 x_1 \leqslant$
2. Equality holds if $x_1 = x_2 = \pi$. If $n \geqslant 3$, fix x_3, x_4, $\ldots$, x_n, then
$x_1 + x_2$ is constant, consider $A = \sin^2 x_1 + \sin^2 x_2$, then

$$2 - 2A = 1 - 2\sin^2 x_1 + 1 - 2\sin^2 x_2 = \cos^2 x_1 + \cos^2 x_2$$
$$= 2\cos(x_1 + x_2)\cos(x_1 - x_2).$$

To find polishing tool, we consider the extremum of $\cos(x_1 - x_2)$
and the signs of $\cos(x_1 + x_2)$. Notice that $\cos(x_1 + x_2) \geqslant 0$ if $x_1 + x_2 \leqslant \frac{\pi}{2}$,
and $\cos(x_1 + x_2) < 0$ if $x_1 + x_2 > \frac{\pi}{2}$. Thus, if $x_1 + x_2 \leqslant \frac{\pi}{2}$, then when
$|x_1 - x_2|$ increases, A will also increase, so we choose the polishing
tool to be $(x_1, x_2) \rightarrow (x_1 + x_2, 0)$; if $x_1 + x_2 > \frac{\pi}{2}$, then when $|x_1 -$
$x_2|$ decreases, A will increase, so the polishing tool is $(x_1, x_2) \rightarrow$
$\left(\frac{x_1 + x_2}{2}, \frac{x_1 + x_2}{2}\right)$. In order to ensure there exist x_1, x_2, such that $x_1 +$
$x_2 \leqslant \frac{\pi}{2}$, a sufficient condition is that $n \geqslant 4$. Thus if $n \geqslant 4$, we can use
polishing transformation as follows. First we prove a lemma: If $0 \leqslant$

x_1, $x_2 \leqslant \frac{\pi}{2}$, and $x_1 + x_2 \leqslant \frac{\pi}{2}$, then $\sin^2 x_1 + \sin^2 x_2 \leqslant \sin^2 (x_1 + x_2)$.

In fact, since $0 \leqslant x_1$, $x_2 \leqslant \frac{\pi}{2}$, $x_1 + x_2 \leqslant \frac{\pi}{2}$, so $| x_1 - x_2 | \leqslant | x_1 + x_2 | \leqslant \frac{\pi}{2}$. Therefore $\cos(x_1 - x_2) \geqslant \cos(x_1 + x_2)$, which implies

$$2 - 2(\sin^2 x_1 + \sin^2 x_2)$$
$$= \cos^2 x_1 + \cos^2 x_2 = 2\cos(x_1 + x_2)\cos(x_1 - x_2)$$
$$\geqslant 2\cos(x_1 + x_2)\cos(x_1 + x_2) = 2\cos^2(x_1 + x_2)$$
$$= 2[1 - \sin^2(x_1 + x_2)] = 2 - 2\sin^2(x_1 + x_2),$$

this proves the lemma. Now if $n = 3$, and the three angles are $\left(\frac{\pi}{2}, \frac{\pi}{2}, 0 \right)$, then we can adjust them to $\left(\frac{\pi}{2}, \frac{\pi}{4}, \frac{\pi}{4} \right)$. The value of F increases from 2 to $1 + \sqrt{2}$, thus we may assume that $x_1 \leqslant x_2 \leqslant x_3$, and $(x_1, x_2, x_3) \neq \left(0, \frac{\pi}{2}, \frac{\pi}{2} \right)$. Then $x_2 < \frac{\pi}{2}$, $x_1 + x_3 > \frac{\pi}{2}$, $x_1 \leqslant \frac{\pi}{3} \leqslant x_3$. Let (x_1, x_2, x_3) be transformed to $\left(\frac{\pi}{3}, x_2, x_1 + x_3 - \frac{\pi}{3} \right)$, from the above discussion, we can see that F increases. Apply polishing transform once again, we get $\left(\frac{\pi}{3}, \frac{\pi}{3}, \frac{\pi}{3} \right)$. Therefore $F \leqslant \frac{9}{4}$. When $n \geqslant 4$, assume that $x_1 \geqslant x_2 \geqslant \cdots \geqslant x_{n-1} \geqslant x_n$, then there must exist two angles such that $x_{n-1} + x_n \leqslant \frac{\pi}{2}$. By Lemma, $F = \sin^2 x_1 + \cdots + \sin^2 x_{n-1} + \sin^2 x_n \geqslant \sin^2 x_1 + \sin^2 x_2 + \cdots + \sin^2 (x_{n-1} + x_n) = \sin^2 x'_1 + \cdots + \sin^2 x'_{n-2} + \sin^2 x'_{n-1}$ (where x'_1, x'_2, $\ldots$, x'_{n-2}, x'_{n-1} is a rearrangement of $x_1, x_2, \ldots, x_{n-2}, x_{n-1} + x_n$ in descending order). If $n - 1 \geqslant 4$, then there must exist two angles such that $x'_{n-2} + x'_{n-1} \leqslant \frac{\pi}{2}$. Employ the lemma again to continue the above transformation, and after a most $n - 3$ transformations, we will obtain $(x'_1, x'_2, x'_3, 0, 0, \ldots, 0)$. Then using the results about $n = 3$, we obtain that $F \leqslant \frac{9}{4}$, where equality holds if $x_1 = x_2 = x_3 = \frac{\pi}{3}$, and $x_4 = x_5 = \cdots = x_n = 0$. In summary, if $n = 2$, then the maximum value of F is 2; if $n > 2$, then the maximum value of F is $\frac{9}{4}$.

(4) Using the method from the preceding example, we can obtain that the maximum value of $\sin a_1 + \sin a_2 + \cdots + \sin a_n$ is $n \sin \frac{A}{n}$.

(5) First, notice that if $x_1 + x_2 = a$ (constant), then since $2 \sin x_1 \sin x_2 = \cos(x_1 - x_2) - \cos a$ and $| x_1 - x_2 | < \pi$, we know that the value of $\sin x_1 \sin x_2$ will increase when $| x_1 - x_2 |$ decreases (polishing tool). If x_1, x_2, x_3, x_4 are not all equal, there must be a number greater than $\frac{\pi}{4}$, and also a number less than $\frac{\pi}{4}$.

Assume that $x_1 > \frac{\pi}{4} > x_2$, fix x_3, x_4, we will to use polishing transform for x_1, x_2; i.e. put $x_1' = \frac{\pi}{4}$, $x_2' = x_1 + x_2 - \frac{\pi}{4}$, $x_3' = x_3$, $x_4' = x_4$, so that $x_1' + x_2' = x_1 + x_2$, $| x_1' - x_2' | < | x_1 - x_2 |$, and $\sin x_1 \sin x_2 < \sin x_1' \sin x_2'$ from above, hence $\sin^2 x_1 \sin^2 x_2 < \sin^2 x_1' \sin^2 x_2'$.

Define $f(x, y) = \left(2 \sin^2 x + \frac{1}{\sin^2 x} \right) \left(2 \sin^2 y + \frac{1}{\sin^2 y} \right)$, then

$$
\begin{aligned}
f(x_1, x_2) &= \left(2 \sin^2 x_1 + \frac{1}{\sin^2 x_1} \right) \left(2 \sin^2 x_2 + \frac{1}{\sin^2 x_2} \right) \\
&= 2 \left(2 \sin^2 x_1 \sin^2 x_2 + \frac{1}{2 \sin^2 x_1 \sin^2 x_2} \right) + 2 \left(\frac{\sin^2 x_1}{\sin^2 x_2} + \frac{\sin^2 x_2}{\sin^2 x_1} \right) \\
&= 2 \left(2 \sin^2 x_1 \sin^2 x_2 + \frac{1}{2 \sin^2 x_1 \sin^2 x_2} \right) + 2 \left(\frac{\sin^4 x_1 + \sin^4 x_2}{\sin^2 x_1 \sin^2 x_2} \right) \\
&= 2 g (2 \sin^2 x_1 \sin^2 x_2) + 2 \left(\frac{\sin^4 x_1 + \sin^4 x_2}{\sin^2 x_1 \sin^2 x_2} \right),
\end{aligned}
$$

where $g(x) = x + \frac{1}{x}$.

Notice that $g(x)$ is decreasing in the $(0, 1)$, and $2 \sin^2 x_1 \sin^2 x_2 \leqslant$

$2\sin^2 x_2 < 2\sin^2\frac{\pi}{4} = 1$, thus

$$g(2\sin^2 x_1 \sin^2 x_2) > g(2\sin^2 x_1' \sin^2 x_2').$$

Therefore,

$$\begin{aligned}
f(x_1, x_2) &= 2g(2\sin^2 x_1 \sin^2 x_2) + 2\left(\frac{\sin^2 x_1}{\sin^2 x_2} + \frac{\sin^2 x_2}{\sin^2 x_1}\right)\\
&> 2g(2\sin^2 x_1' \sin^2 x_2') + 2\left(\frac{\sin^4 x_1' + \sin^4 x_2'}{\sin^2 x_1' \sin^2 x_2'}\right)\\
&= f(x_1', x_2'),
\end{aligned}$$

$$\begin{aligned}
A(x_1, x_2, x_3, x_4) &= f(x_1, x_2)f(x_3, x_4)\\
&> f(x_1', x_2')\, f(x_3', x_4')\\
&= A(x_1', x_2', x_3', x_4').
\end{aligned}$$

Thus, after transforming x_1 to $\frac{\pi}{4}$, the function value decreases. If there exists a number in x_2', x_3', x_4' which is not equal to $\frac{\pi}{4}$, then we can continue the above-mentioned transform. This shows that A reaches the minimum when $x_1 = x_2 = x_3 = x_4 = \frac{\pi}{4}$.

In summary, the minimum value of A is $\left(2\sin^2\frac{\pi}{4} + \dfrac{1}{\sin^2\frac{\pi}{4}}\right)^4 = 81$.

Exercise 6

(1) $|X|_{\min} = 9$. Denote $X = \{a_1, a_2, \ldots, a_n\}$, where $1 = a_1 < a_2 < \cdots < a_n = 100$. We estimate the spacing between a_k and a_{k-1}. Take any $a_k \in X(k > 1)$, there exist a_i, a_j, such that $a_k = a_i + a_j$, where $a_i < a_k$, $a_j < a_k$, so $a_i \leqslant a_{k-1}$, $a_j \leqslant a_{k-1}$, thus $a_k = a_i + a_j \leqslant a_{k-1} + a_{k-1} = 2a_{k-1}$. If $a_k = 2a_{k-1}$ for all k, then the numbers in X are powers of 2, which contradicts $100 \in X$. Thus there exists a number k, such that $a_k < 2a_{k-1} = a_{k-1} + a_{k-1}$, which implies

$$a_k \leqslant a_{k-1} + a_{k-2} \leqslant 2a_{k-2} + a_{k-2} = 3a_{k-2}.$$

Using this k, we obtain that

$$100 = a_n \leqslant 2a_{n-1} \leqslant 4a_{n-2} \leqslant \cdots \leqslant 2^{n-k}a_k \leqslant 3 \times 2^{n-k}a_{k-2}$$
$$\leqslant 3 \times 2^{n-k+1}a_{k-3} \leqslant \cdots \leqslant 3 \times 2^{n-3}a_1 = 3 \times 2^{n-3},$$

therefore $n \geqslant 9$. On the other hand, when $n = 9$, there exists a set satisfying the condition of the problem: $X = \{100, 50, 25, 13, 12, 6, 3, 2, 1\}$, so the minimum value of n is 9.

Another solution: Denote $X = \{a_1, a_2, \ldots, a_n\}$, where $1 = a_1 < a_2 < \cdots < a_n = 100$. We estimate the spacing between a_k and a_{k-1}. Take any $a_k \in X (k > 1)$, there exist a_i, a_j, such that $a_k = a_i + a_j$, where $a_i < a_k$, $a_j < a_k$, so $a_i \leqslant a_{k-1}$, $a_j \leqslant a_{k-1}$, thus $a_k = a_i + a_j \leqslant a_{k-1} + a_{k-1} = 2a_{k-1}$. It implies that $100 = a_n \leqslant 2a_{n-1} \leqslant 4a_{n-2} \leqslant \cdots \leqslant 2^{n-1}a_1 = 2^{n-1}$, therefore $n \geqslant 8$. If $n = 8$, then $1 = a_1 < a_2 < \cdots < a_8 = 100$. It is easy to know that $a_2 = 2 = 2^{2-1}$, otherwise a_2 cannot be represented as the sum of two numbers in X. Then notice that $a_8 = 100$ is not a power of 2, we may assume that there exists $a_t \in X$ such that $a_t \neq 2^{t-1}$, and $a_k = 2^{k-1}$ for all $k = 1, 2, \ldots, t-1$. Since a_t can be represented as the sum of two numbers in X, so there exist $a_i, a_j \in X(i, j < t)$, such that $a_t = a_i + a_j$. If $a_i = a_j$, then $a_t = 2a_i = 2 \times 2^{i-1} = 2^i$, which is impossible. Hence $a_i \neq a_j$, then $a_j < 2^{t-2}$, $a_i < 2^{t-3}$, so that $a_t = a_i + a_j < 2^{t-2} + 2^{t-3}$. therefore $100 = a_8 = 2^{8-t}a_t < 2^{8-t}(2^{t-2} + 2^{t-3}) = 2^6 + 2^5 = 96$, which is a contradiction. Thus $n \neq 8$, hence $n \geqslant 9$.

(2) $C_g = n+1$. First of all, we may fill ith row with $(i-1)n+1$, $(i-1)n+2, \ldots, (i-1)n+n$ in that order, so that the c-gap of the number table is $n+1$. Now for any number table, let g be the c-gap of the table. Then $|x-y| \leqslant g$ for any two connected numbers x, y. We will prove that $g \geqslant n+1$. Connect by a chain the two grids filled with 1 and n^2 respectively, so that the chain contains at most n grids including the end-points. We may assume there are m grids $(m \leqslant n)$,

and the numbers in the grids are $a_1 = 1$, a_2, a_3, ..., $a_m = n^2$. Examining the difference of each pair of adjacent numbers, we have

$$|a_2 - a_1| + |a_3 - a_2| + \cdots + |a_m - a_{m-1}|$$
$$\geqslant (a_2 - a_1) + (a_3 - a_2) + \cdots + (a_m - a_{m-1})$$
$$= a_m - a_1 = n^2 - 1.$$

Therefore, there must exist i , such that

$$|a_i - a_{i-1}| \geqslant \frac{n^2 - 1}{m - 1} \geqslant \frac{n^2 - 1}{n - 1} = n + 1.$$

(3) Consider the general problem. Let $f(n)$ be the maximum of the number of self-intersection points of a closed path composed of n line segments (n is odd). We fix a vertex, walk along the path starting from this point, and label the segments 1, 2, ..., n , respectively. It is clear that the ith segment intersects at most $i - 2$ previous segments, where $i = 1, 2, ..., n - 1$. Finally, the nth segments intersects at most $n - 3$ previous segments, so the number of the intersection points is not more than

$$\sum_{i=3}^{n-1} (i - 2) + (n - 3) = \frac{1}{2}(n - 2)(n - 3) + \frac{1}{2} \cdot 2(n - 3) = \frac{1}{2}n(n - 3).$$

Therefore $f(n) \leqslant \frac{1}{2}n(n - 3)$. On the other hand, when n is odd, there exists a closed path with n segments such that $f(n) = \frac{1}{2}n(n - 3)$. In fact, choose two points A_1 and A_2, draw a circle with A_1A_2 being a diameter, and locate the points A_1, A_3, ..., A_n on the circle, and arrange them in the counterclockwise direction. Similarly, locate the points A_2, A_4, A_6, ..., A_{n-1} on the circle and arrange them in a counterclockwise direction. There are then $\frac{1}{2}n(n - 3)$ self-intersection points. In summary, the answer of this problem is

$$f(2005) = \frac{1}{2}(2005 \times 2002) = 2005 \times 1001 = 2\,007\,005.$$

(4) Denote the number of occurrences of i, $i = 1, 2, \ldots, 10$ in the sequence $A_1, A_2, \ldots, A_k$ by d_i. We first prove that $d_i \leqslant 4$, $i = 1, 2, \ldots, 10$. In fact, for any $i \in M$, the ordered 2-tuples composed of the element " i " and the elements " j " which belongs to the other nine elements in M, occurs at most twice in $A_1, A_2, \ldots, A_k$ which are the 5-subset satisfying the conditions of the problem. Therefore the total number of appearances of i in the nine 2-tuples i, j is at most 18, thus $4 \cdot d_i \leqslant 18$ and hence $d_i \leqslant 4$. In addition, there are $5k$ elements in k 5-element subsets and the number of each element of M appearing in $A_1, \ldots, A_k$ is d_i, so $5k = d_1 + \cdots + d_{10} \leqslant 4 \times 10$, hence $k \leqslant 8$. Finally, suppose $k = 8$, the following eight 5-element sets of numbers satisfy the requirements: $A_1 = \{1, 2, 3, 4, 5\}$, $A_2 = \{1, 6, 7, 8, 9\}$, $A_3 = \{1, 3, 5, 6, 8\}$, $A_4 = \{1, 2, 4, 7, 9\}$, $A_5 = \{2, 3, 6, 7, 10\}$, $A_6 = \{3, 4, 7, 8, 10\}$, $A_7 = \{4, 5, 8, 9, 10\}$, $A_8 = \{2, 5, 6, 9, 10\}$. Thus $k_{\min} = 8$.

Exercise 7

(1) Divide X into 47 subsets, $A_i = \{x \mid x \equiv i, \text{ or } x \equiv 93 - i \ (\text{mod } 93)\}$, $i = 0, 1, 2, \ldots, 46$. Whenever $x, y \in A_i$, we can get that $x - y \equiv 0 \ (\text{mod } 93)$, or $x + y \equiv 0 \ (\text{mod } 93)$, so in A there is at most one number that is contained in A_i, so $|A| \leqslant 47$. Let $A = \{10, 11, \ldots, 46\} \cup \{93, 94, \ldots, 101\} \cup \{84\}$, then A meets the requirement, so the maximum of $|A|$ is 47.

(2) First, let $A_1 = \{1, 2, 3, 5, 8\}$, $A_2 = \{4, 6, 10\}$, $A_3 = \{7, 9\}$, so that any three numbers in A_i shall not be the side lengths of a triangle, so in A there are at most two numbers that are contained in A_i, so $|A| \leqslant 2 \times 3 = 6$. In addition, let $A = \{5, 6, 7, 8, 9, 10\}$, then A meets the requirement, so the maximum of $|A|$ is 6.

(3) First, let $A_1 = \{1, 2, 3, 5, 8, 13\}$, $A_2 = \{4, 6, 10, 16\}$, $A_3 = \{7, 12, 19\}$, $A_4 = \{9, 11, 20\}$, $A_5 = \{14, 15\}$, $A_6 = \{17, 18\}$, so that any three numbers in A_i shall not be the side lengths of a triangle, so in A there are at most two numbers that are contained in A_i, so $|A| \leqslant 2 \times 6 = 12$. If $|A| = 12$, then $14, 15, 17, 18 \in A$, then $1, 2, 3, 5 \notin A$, so $8, 13 \in A$ and $7, 9 \notin A$, therefore $12, 19, 11, 20 \in A$, but now $8 + 12 = 20$, which is a contradiction. So $|A| \leqslant 11$. In addition, let $A = \{10, 11, 12, \ldots, 20\}$, then A satisfies the requirement, so the maximum of $|A|$ is 11.

(4) Let $A_{ij} = \{\overline{ij}, \overline{ji}\}$, $i, j \in \{0, 1, 2, \ldots, 9\}$, then A contains at least one element in A_{ij}, otherwise in the infinite sequence $ijijijij\ldots$ there will be no adjacent number pair belonging to A. Obviously there are $10 + C_{10}^2 = 55$ sets in A_{ij} in total, ten of which are A_{00}, A_{11}, $\ldots$, A_{99}, so $|A| \geqslant 55$. In addition, let $A = \{\overline{ij} \mid 0 \leqslant i \leqslant j \leqslant 9\}$, namely all the numbers in A_{ij} should be such that $i \leqslant j$, and then $|A| = 55$. Now for any infinite sequence, suppose its minimum number is i and the one following it is j, then $i \leqslant j$ and $\overline{ij} \in A$. Thus the minimum of $|A|$ is 55.

(5) Let $A_1 = \{1, 2, 4, 8, 16\}$, $A_2 = \{3, 6, 12\}$, $A_3 = \{5, 10, 20\}$, $A_4 = \{7, 14\}$, $A_5 = \{9, 18\}$, $A_6 = \{11\}$, $A_7 = \{13\}$, $A_8 = \{15\}$, $A_9 = \{17\}$, $A_{10} = \{19\}$. Since in the three sets $\{1, 2\}$, $\{4, 8\}$ and $\{16\}$, we can take at most one number from each, we know that in A_1 we can take at most three numbers. If we take three numbers from A_1, the only possibility is 16, 4 and 1. Similarly, in A_2 we can take at most two numbers, and there is only one way to take two numbers. Now in both in A_4 and A_5 we can take only one number and for each there are two ways to take one number. As for other sets, we can take only one number and there is only one way to take one number. As a result, the total number that we can take is $3 + 2 + 2 + (1 + 1) + 1 \times 5 = 14$ and equality can hold. The number of ways in which we take the numbers is $2 \cdot 2 = 4$. The set of these taken numbers is $X \cup Y_i$ $(i = 1,$

2, 3, 4), where $X = \{1, 4, 16, 3, 12, 5, 20, 11, 13, 15, 17, 19\}$, $Y_1 = \{7, 8\}$, $Y_2 = \{7, 9\}$, $Y_3 = \{14, 8\}$, $Y_4 = \{14, 9\}$, so the maximum number of the taken numbers is 14.

Alternative solution: let $A_1 = \{1, 2\}$, $A_2 = \{3, 6\}$, $A_3 = \{4, 8\}$, $A_4 = \{5, 10\}$, $A_5 = \{7, 14\}$, $A_6 = \{9, 18\}$. From each set we take only one number, thus we can take at most six numbers from these sets. Besides these sets there are eight numbers, thus the number of the numbers we can take is $6 + 8 = 14$. If we take 14 numbers, these must include 11, 12, 13, 15, 16, 17, 19, 20. Notice that we cannot take 6, 8 and 10 if we take 12, 16 and 20, and that we can 3, 4 and 5. Also if we take the number of 4, then we cannot take 2, and also we can take 1. Also, since we take one number from each of A_5 and A_6, there are $2 \times 2 = 4$ way to take the numbers.

(6) The maximum of K is 663. First prove that $k \leqslant 663$. Note that when $x - y = 1$ or 2, we have $x - y \mid x + y$, thus we can conclude that if we take any two numbers x and y in any three consecutive natural numbers, we must have $x - y \mid x + y$. When $k > 663$, let $A_i = \{3i - 2, 3i - 1, 3i\}(i = 1, 2, \ldots, 662)$, $A_{663} = \{1987, 1988\}$, then at least one set contains two chosen numbers x and y, so $x - y \mid x + y$, which is a contradiction. Besides, when $k = 663$, let $A = \{1, 4, 7, \ldots, 1987\}$, then for any two numbers a_i and a_j in A, we can conclude that $a_i - a_j = (3i - 2) - (3j - 2) = 3(i - j)$, and $a_i + a_j = (3i - 2) + (3j - 2) = 3(i + j - 1) - 1$, thus $a_i + a_j$ cannot be exactly divided by $a_i - a_j$.

(7) Divide X into two subsets, $A_1 = \{x \mid x \equiv 0 \pmod{7}, x \in X\}$, $A_2 = X \backslash A_1$, then $|A_1| = 7$, $|A_2| = 43$. Obviously, in S there is only one number contained in A_1, so $|S| \leqslant 43 + 1 = 44$. On the other hand, for any integer x, if $x \equiv 0, \pm 1, \pm 2, \pm 3 \pmod 7$, then $x^2 \equiv 0, 1, 4, 2 \pmod 7$, from which we can see that if $x^2 + y^2 \equiv 0 \pmod 7$, then $x^2 \equiv y^2 \equiv 0 \pmod 7$. Thus let $S = A_2 \cup \{7\}$, then S satisfies the requirement and $|S| = 44$, so the maximum of $|S|$ is 44.

(8) Let $A_k = \{k, 19k\}$, $k = 6, 7, \ldots, 105$, then in A there is only one number contained in A_k, so $|A| \leqslant 1995 - 100 = 1895$. On the other hand, let $A = \{1, 2, 3, 4, 5\} \cup \{106, 107, \ldots, 1995\}$, then A satisfies the requirement and $|A| = 1895$, so the maximum of $|A|$ is 1895.

Exercise 8

(1) The minimum of n is 27. First prove $n \geqslant 27$. If $n \leqslant 26$, let $A = \{1, 3, 5, \ldots, 2n-1\}$, then $|A| = n$, but the sum of any two numbers in A is $(2i-1) + (2j-1) = 2(i+j) - 2 \leqslant 2(26+25) - 2 = 100 < 102$ (where $0 < i < j \leqslant n \leqslant 26$), which is a contradiction. Now, when $n = 27$, let $A_i = \{2i+1, 101-2i\}(i = 1, 2, \ldots, 24)$, $A_{25} = \{1\}$, $A_{26} = \{51\}$. For any 27 numbers, there must be two numbers belonging to the same set, so their sum is 102.

(2) Let $A = \{1, 2, \ldots, 8\} \cup \{134, 135, \ldots, 1995\}$, then obviously A satisfies the conditions, and $|A| = 1870$. On the other hand, considering 125 sets where $A_k = \{k, 15k\}(k = 9, 10, \ldots, 133)$ we have 250 different numbers and if we exclude these numbers, there are still 1745 numbers in X. Now let us make everyone of these 1745 numbers into a singleton set, so that together with the former 125 sets, we have $1745 + 125 = 1870$ sets in total. If $|A| > 1870$, then in A there must be two numbers contained in the same set, so the bigger number is 15 times the smaller one, which is a contradiction. Thus the maximum of $|A|$ is 1870.

(3) Suppose that F satisfies the conditions. If we want to maximize $|F|$, notice that $|A_i \cap A_j| \leqslant 2$, we should take some sets that have fewer elements and include them in F. Obviously, all the sets with at most two elements can be included in F. In addition, all the sets with at most three elements can also be included in F. In fact, for any two sets A_i and A_j in F, when one of them has at most two

elements, then clearly $| A_i \cap A_j | \leqslant 2$; when $| A_i | = | A_j | = 3$ and $| A_i \cap A_j | > 2$, we know $| A_i \cap A_j | = 3$, thus $A_i = A_j$, which is a contradiction. Thus we can get that

$$| F |_{\max} \geqslant C_{10}^1 + C_{10}^2 + C_{10}^3 = 175.$$

The above set F is saturated in that we cannot put in any other set, but it does not mean $| F |$ is the maximum. The next step is to prove that $| F | \leqslant 175$. Now assume that $| F | > 175$, then there must be a set A such that $| A | > 3$. Choose any one 3-element subset from A, and call it A'. Because $| A \cap A' | = 3 > 2$ and A belongs to F, we know that A' does not belong to F. Now we replace A by A' to obtain F', then F' also satisfies the conditions. In this way, we can change all the sets with more than three elements into 3-element subsets. Now $| F | = | F^* | \leqslant 175$, which is a contradiction.

(4) Let $A_i = \{i, i+1, i+2, \ldots, i+59\}$ $(i = 1, 2, \ldots, 70)$, and take the elements modulo 70 (namely: when $x > 70$, we replace x by $x - 70$). Obviously $60 \in A_1, A_2, \ldots, A_{60}$, so $k_{\max} \geqslant 60$. Now we prove that $k_{\max} = 60$. Assume $k \geqslant 61$, we have to show that for any k sets we choose, there are always 7 of them whose intersection is empty set. The idea is to make the union of their complementary sets be the whole set, namely that $\overline{A_{i_1}} \cup \overline{A_{i_2}} \cup \ldots \cup \overline{A_{i_7}} = I$. Notice that $\overline{A_{i_1}}$, $\overline{A_{i_2}}, \ldots, \overline{A_{i_7}}$ are all 10-element subsets; since there are 70 elements in I, we need that these seven subsets have the form of $A_i, A_{10+i}, \ldots, A_{60+i}$. Now the problem is to find an I such that all the seven sets are chosen. Note that the subscripts of all these sets are congruent to I modulo 10. After choosing $k \geqslant 61$ sets, there are at most nine sets left, and their subscripts cannot cover all residue classes modulo 10; thus there exists $0 \leqslant i \leqslant 9$, so that the sets $A_i, A_{10+i}, \ldots, A_{60+i}$ are all chosen. Since $A_i \cap A_{10+i} \cap \cdots \cap A_{60+i} = \varPhi$, we have a contradiction.

(5) The maximum number of elements in S is 76. Suppose $| S | \geqslant 3$, $p_1^{q_1} p_2^{q_2} p_3^{q_3} \in S$, and p_1, p_2 and p_3 are three different prime

numbers, and α_1, α_2 and α_3 are positive integers. Also suppose $q \in \{2, 3, 5, 7\}$, $q \neq p_1$, p_2, p_3, so $\{p_1, p_2, p_3, q\} = \{2, 3, 5, 7\}$. From (i) we know that there exists $c_1 \in S$ such that $(p_1^{\alpha_1} p_2^{\alpha_2} p_3^{\alpha_3}, c_1) = 1$; choose c_1 to have the minimum minimal prime factor.

From (i) we can conclude that there exists $c_2 \in S$, such that $(c_2, c_1) = 1$, $(c_2, p_1^{\alpha_1} p_2^{\alpha_2} p_3^{\alpha_3}) = 1$.

From (ii) we know that there exists $c_3 \in S$, such that $(c_3, c_1) > 1$, $(c_3, c_2) > 1$. From $(c_1, c_2) = 1$ we can see that the product of the minimum prime factors of c_1 and c_2 is less than or equal to $c_3 \leqslant 108$, therefore $q \mid c_1$. From $(c_2, c_1) = 1$, $(c_2, p_1^{\alpha_1} p_2^{\alpha_2} p_3^{\alpha_3}) = 1$, $\{p_1, p_2, p_3, q\} = \{2, 3, 5, 7\}$ and $c_2 \leqslant 108$, we know that c_2 is a prime number larger than 10. From $(c_3, c_2) > 1$ we know that $c_2 \mid c_3$. Also $1 < \left(c_1, \dfrac{c_3}{c_2}\right) < 10$, $(p_1^{\alpha_1} p_2^{\alpha_2} p_3^{\alpha_3}, c_1) = 1$, so $\left(c_1, \dfrac{c_3}{c_2}\right) = q^{\alpha}$ ($*$). From (i) we can conclude that there exists $c_4 \in S$, such that $(c_4, p_1^{\alpha_1} p_2^{\alpha_2} p_3^{\alpha_3}) = 1$, $(c_4, c_3) = 1$, so from($*$)and $(c_4, p_1 p_2 p_3 q) = 1$ (which is just $(c_4, 2 \times 3 \times 5 \times 7) = 1$), we can conclude that c_4 is a prime number larger than 10. By (ii) we know that there exists $c_5 \in S$, such that $(c_5, c_2) > 1$, $(c_5, c_4) > 1$, so $c_2 \mid c_5$, $c_4 \mid c_5$. Also $c_2 \mid c_3$, $(c_4, c_3) = 1$, thus $(c_2, c_4) = 1$, thus $c_2 c_4 \mid c_5$. But $c_2 c_4 \geqslant 11 \times 13 > 108$, which is a contradiction.

Now let $S_1 = \{1, 2, \ldots, 108\} \setminus (\{1$ and prime numbers more than $11\} \cup \{2 \times 3 \times 11, 2 \times 3 \times 5, 2^2 \times 3 \times 5, 2 \times 3^2 \times 5, 2 \times 3 \times 7, 2^2 \times 3 \times 7, 2 \times 5 \times 7, 3 \times 5 \times 7 \})$, then $|S_1| = 76$. Now we check that S_1 satisfies (i), (ii). If $p_1^{\alpha_1} p_2^{\alpha_2} p_3^{\alpha_3} \in S_1$, we have $p_1 < p_2 < p_3$, and $p_3 \geqslant 11$. So $p_1^{\alpha_1} p_2^{\alpha_2} p_3^{\alpha_3} = 2 \times 3 \times 13$ or $2 \times 3 \times 17$.

① $a = 2 \times 3 \times 13$, $b \in S_1$, $b \neq a$. (a) Since there is at least one number p among 5, 7 and 11 that does not divide b, we know $(a, p) = (b, p) = 1$. (b) Suppose that the minimum prime factor of b is q_1, then $2q_1 \leqslant 108$, $3q_1 \leqslant 108$. If $b \neq 2q_1$, then $2q_1 \in S_1$, $(2q_1, a) > 1$, $(2q_1, b) > 1$; if $b \neq 3q_1$, then $3q_1 \in S_1$, $(3q_1, a) > 1$, $(3q_1, b) > 1$.

② $a = 2 \times 3 \times 17$, $b \neq a$, so the proof goes the same way as in ①.

③ $a = b$. Since there is at least one number among 5, 7 and 11 that cannot divide a, we know that (i) is true. If a is composite, denote its minimum prime factor by p, then $p \in S_1$, $(p, a) > 1$; if a is prime, then $a \leqslant 11$, $2a \in S_1$, $(2a, a) > 1$.

④ Suppose that a and b are two different numbers of S_1, and both a and b contain at most two different prime factors, and $a < b$. (a) We know that there is a p among 2, 3, 5, 7 and 11 that does not divide ab, and $p \in S_1$, $(p, a) = (p, b) = 1$. (b) Suppose that the minimum prime factors of a and b are respectively r_1, r_2, then $r_1 r_2 \leqslant 108$. If $r_1 = r_2 < a$, then $r_1 \in S_1$, $(a, r_1) > 1$, $(b, r_1) > 1$; if $r_1 = r_2 = a$, we choose $u = 2$ or 3, such that $b \neq ua$. Thus $ua \in S$, $(ua, a) > 1$, $(ua, b) > 1$. If $r_1 r_2 \neq a$, $r_1 r_2 \neq b$, then $r_1 r_2 \in S_1$, $(r_1 r_2, a) > 1$, $(r_1 r_2, b) > 1$. If $r_1 r_2 = a$, then $r_1 < r_2$, we choose $u = 2, 3, 5$, so that $b \neq ur_2$, $a \neq ur_2$, then $ur_2 \in S_1$, $(ur_2, a) > 1$, $(ur_2, b) > 1$. If $r_1 r_2 = b$, then we choose $v = 2, 3, 5$, so that $a \neq ur_1$, $b \neq ur_1$, thus $vr_1 \in S_1$, $(vr_1, a) > 1$, $(vr_1, b) > 1$. We have verified that S_i satisfied (i), (ii).

On the other hand, we have seen that $2 \times 3 \times 5$, $2^2 \times 3 \times 5$, $2 \times 3^2 \times 5$, $2 \times 3 \times 7$, $2^2 \times 3 \times 7$, $2 \times 5 \times 7$, $3 \times 5 \times 7$ do not belong to S. Now let us verify that $2 \times 3 \times 11$, $2 \times 3 \times 13$, 5×7 cannot belong to S simultaneously. Assume the contrary; from (i) we know that there exist d_1, $d_2 \in S$, such that $(2 \times 3 \times 11, d_1) = 1$, $(5 \times 7, d_1) = 1$, $(2 \times 3 \times 13, d_2) = 1$, $(5 \times 7, d_2) = 1$, thus d_1 and d_2 are all prime numbers larger than 10. From (ii) we know that $d_1 = d_2 \geqslant 17$. From (ii) we know that there exists $d_3 \in S$, such that $(7, d_3) > 1$, $(d_2, d_3) > 1$, so $7d_2 \mid d_3$. However, $7d_2 \geqslant 7 \times 17 = 119$, which is a contradiction. On the other hand, there is at most one number among the prime numbers larger than 10 that belongs to S, and $1 \notin S$, thus $|S| \leqslant 108 - 7 - 1 - 23 - 1 = 76$.

(6) First, for $n \leqslant 16$, we shall construct a coloring method such that it does not satisfy the requirement of the problem.

Let A_1, A_2, $\ldots$, A_n be the vertices of the regular n-gon in counterclockwise order, and M_1, M_2, M_3 be the sets of vertices with the three colors respectively.

When $n = 16$, let $M_1 = \{A_5, A_8, A_{13}, A_{14}, A_{16}\}$, $M_2 = \{A_3, A_6, A_7, A_{11}, A_{15}\}$, $M_3 = \{A_1, A_2, A_4, A_9, A_{10}, A_{12}\}$. For M_1, the distances from A_{14} to the other four vertices are different from each other while the other four vertices form a rectangle. Similarly for M_2, we can verify that there does not exist four vertices forming some isosceles trapezoid. As for M_3, the 6 it happens to contain the six vertices from three diameters, thus any four vertices also cannot form an isosceles trapezoid.

When $n = 15$, let $M_1 = \{A_1, A_2, A_3, A_5, A_8\}$, $M_2 = \{A_6, A_9, A_{13}, A_{14}, A_{15}\}$, $M_3 = \{A_4, A_7, A_{10}, A_{11}, A_{12}\}$, we see that it is impossible for the four vertices in any M_i to form an isosceles trapezoid.

When $n = 14$, let $M_1 = \{A_1, A_3, A_8, A_{10}, A_{14}\}$, $M_2 = \{A_4, A_6, A_7, A_{11}, A_{12}\}$, $M_3 = \{A_2, A_6, A_9, A_{13}\}$, we see that it is impossible for the four vertices in any M_i to form an isosceles trapezoid.

When $n = 13$, let $M_1 = \{A_5, A_6, A_7, A_{10}\}$, $M_2 = \{A_1, A_8, A_{11}, A_{12}\}$, $M_3 = \{A_2, A_3, A_4, A_9, A_{13}\}$, we see that it is impossible for the four vertices in any M_i to form an isosceles trapezoid.

In the above-mentioned situation, if we delete the vertex A_{13}, we can get the coloring method of $n = 12$, and then we may delete the vertex A_{12} to get the coloring method of $n = 11$, and delete the vertex A_{11} to get the coloring method of $n = 10$.

When $n \leqslant 9$, it is possible to make the number of vertices with each color to be less than 4, thus there cannot be four vertices with the same color forming an isosceles trapezoid.

In conclusion, the claim is not true if $n \leqslant 16$.

Next let us verify the conclusion is true when $n = 17$.

Assume the contrary, that we can color the vertices of a 17-polygon using three colors, making it impossible for four vertices with the same color to form an isosceles trapezoid.

Since $\left[\frac{17}{3}\right] + 1 = 6$, there must exist six vertices with the same color. We may have the color yellow. If we connect any two of these six vertices, we get $C_6^2 = 15$ segments. Because these segments have only $\left[\frac{17-1}{2}\right] = 8$, possible lengths, we must have one of the two following scenarios:

(i) There exist some three segments that are with the same length.

Note that $3 \nmid 17$, we know that these three segments cannot form a triangle. As a result, there exist two segments connecting four different vertices, thus the four vertices form an isosceles trapezoid, which is a contradiction.

(ii) There are seven pairs of segments with the same length. From assumption, we know that every pair of segments with the same length must have a common yellow vertex, since otherwise we can find four yellow vertices forming an isosceles trapezoid. Now according to the pigeonhole principle, there must be two pairs of segments whose common vertex is the same yellow vertex. In this case, the other vertices of the four segments must form an isosceles trapezoid, which is a contradiction. Thus the claim is true when $n = 17$.

In conclusion, the minimum of n is 17.

Exercise 9

(1) First we simplify the conditions. Let $A_i' = \{$numbers less than 1988 in $A_i\} = A_i \cap \{1, 2, \ldots, 1988\}$, then $N_i(1988) = |A_i'|$, $N_{ij}(1988) = |A_i' \cap A_j'|$, therefore we should show that there exist $i, j (1 \leqslant i < j \leqslant 29)$, such that $|A_i' \cap A_j'| > 200$, where A_i' is the set of numbers in A_i that are 1988, namely $A'_i = A_i \cap \{1, 2, \ldots, 1988\}$. From the assumptions we know that for any i, $|A_i'| = N_i(1988) \geqslant \frac{1988}{e} > 731$, so $|A_i'| \geqslant 732$. Let $X = \{1, 2, 3, \ldots, 1988\}$, then $A_1', A_2', \ldots, A'_{1988}$ are subsets of X. We may as well suppose that $|A'_i| = 732$; otherwise delete some elements in A'_i. Consider the set-element relation table and assume that there are m_i 1s in the ith row; we

calculate the total times that the elements appear in the subsets. Thus we have $\sum_{i=1}^{1988} m_i = S = \sum_{i=1}^{29} |A_i'| = 732 \times 29$. Next we calculate the total times that the elements appear in the intersections, getting $\sum_{i=1}^{1988} C_{m_i}^2 = T = \sum_{1 \leqslant i < j \leqslant 29} |A_i' \cap A_j'|$. Therefore, from the Cauchy-Schwartz inequality, we can get

$$2 \sum_{1 \leqslant i < j \leqslant 29} |A_i' \cap A_j'| = 2 \sum_{i=1}^{1988} C_{m_i}^2 = \sum_{i=1}^{1988} m_i^2 - \sum_{i=1}^{1988} m_i$$

$$\geqslant \frac{\left(\sum_{i=1}^{1988} m_i\right)^2}{\sum_{i=1}^{1988} 1^2} - \sum_{i=1}^{1988} m_i = \frac{(732 \times 1988)^2}{1988} - 732 \times 29,$$

so there must exist one $A_i' \cap A_j'$, such that

$$|A_i' \cap A_j'| \geqslant \frac{\dfrac{(732 \times 1988)^2}{1988} - 732 \times 29}{2C_{29}^2} > 253.$$

Namely, there must be one $A_i' \cap A_j'$, such that $|A_i' \cap A_j'| \geqslant 200$.

(2) Because $\sum_{i=1}^{10} m_i = \sum_{i=1}^{k} 5 = 5k$, from the Cauchy-Schwartz inequality we can get

$$2 \sum_{1 \leqslant i < j \leqslant k} |A_i \cap A_j| = 2 \sum_{i=1}^{10} C_{m_i}^2 = \sum_{i=1}^{10} m_i^2 - \sum_{i=1}^{10} m_i$$

$$\geqslant \frac{\left(\sum_{i=1}^{10} m_i\right)^2}{\sum_{i=1}^{10} 1^2} - \sum_{i=1}^{10} m_i = \frac{25k^2}{10} - 5k.$$

Adding in the condition that $\sum_{1 \leqslant i < j \leqslant k} |A_i \cap A_j| \leqslant \sum_{1 \leqslant i < j \leqslant k} 2 = 2C_k^2 = k^2 - k$ and combing these two inequalities, we get $k \leqslant 6$. When $k = 6$, the six sets $\{1, 2, 3, 4, 5\}$, $\{3, 5, 7, 8, 9\}$, $\{1, 2, 6, 7, 8\}$, $\{1, 3, 6, 9, 10\}$, $\{2, 4, 7, 9, 10\}$, $\{4, 5, 6, 8, 10\}$ meet the requirements, so the maximum of k is 6.

(3) **Solution 1.** We inspect the relation table $B(n, m)$ of m sets and n elements, calculating the number of 1 appearing in this table. Suppose that there are t_i 1s in row i $(i = 1, 2, \ldots, n)$, then $\sum_{i=1}^{n} t_i = S = \sum_{j=1}^{m} |A_j| = rm$. Computing the total times the elements appear in intersections of sets, we get $\sum_{i=1}^{n} C_{t_i}^2 = T = \sum_{1 \leqslant i < j \leqslant m} |A_i \cap A_j|$.

Therefore, from the Cauchy-Schwartz we can get

$$2k C_m^2 = 2 \sum_{1 \leqslant i < j \leqslant m} |A_i \cap A_j| = 2 \sum_{i=1}^{n} C_{t_i}^2 = \sum_{i=1}^{n} t_i^2 - \sum_{i=1}^{n} t_i$$

$$\geqslant \frac{\left(\sum_{i=1}^{n} t_i\right)^2}{\sum_{i=1}^{n} 1^2} - \sum_{i=1}^{n} t_i = \frac{r \cdot m^2}{n} - rm.$$

As a result, $km(m-1) \geqslant \dfrac{r \cdot m^2}{n} - rm$, thus $n \geqslant \dfrac{mr^2}{r + (m-1)k}$.

Solution 2. Suppose that the total time that an element x appears in $A_1, A_2, \ldots, A_m$ is $d(x)$, and call it the degree of x, then

$$\sum_{x, x \in X} d(x) = \sum_{i=1}^{m} |A_i|.$$

For the set A_i, the degree sum of its all elements is called the degree of A_i. Denoting it by $d(A_i)$, we have $d(A_i) = \sum_{x \in A_i} d(x)$. We inspect the degree sum $S = \sum_{i=1}^{m} d(A_i)$ of all the sets. On the one hand,

$$d(A_i) = \sum_{\substack{j \neq i \\ 1 \leqslant j \leqslant m}} |A_i \cap A_j| + |A_i| \leqslant \sum_{\substack{j \neq i \\ 1 \leqslant j \leqslant m}} k + r = (m-1)k + r.$$

Thus $S = \sum_{i=1}^{m} d(A_i) \leqslant \sum_{i=1}^{m} [r + (m-1)k] = m[r + (m-1)k]$. On the other hand, for fixed A_i, when $x \in A_i$, the contribution of x to $d(A_i)$ is $d(x)$. Noting that x appears in $d(x)$ numbers of A_i, we know that the contribution of x to S is $d(x)^2$. Therefore, by the Cauchy-Schwartz, we can get

$$S = \sum_{i=1}^{m} \sum_{x,\, x \in A_i} d(x) = \sum_{x,\, x \in X} d(x)^2 \geqslant \frac{\left(\sum_{x \in X} d(x)\right)^2}{\sum_{x \in X} 1}$$

$$= \frac{\left(\sum_{i=1}^{m} |A_i|\right)^2}{|X|} = \frac{(mr)^2}{|X|},$$

which completes the proof.

(4) Suppose that there are n questions; we shall prove $n_{\max} = 5$. Obviously $n_{\max} > 1$. When $n > 1$, for some question A, if five people choose the same answer, then the answers that the five people choose from any other question B must be different from each other. However, there are only four choices for B, which is a contradiction. Therefore, for any question, there are at most four people making the same choice. Since there are only four choices for A, and that there are 16 people, we know that every answer is chosen by four people. For every person x and each question $i \, (i = 1, 2, \ldots, n)$, there are precisely three people choosing the same answer as x on question i, forming a 3-element subset A_i. Consider the sets $A_1, A_2, \ldots, A_n$; if there are two $A_i, A_j \, (i < j)$ with nonempty intersection, say $y \in A_i \cap A_j$, then the answers of x, y to the two test questions i, j are the same, which is a contradiction. Thus the A_is are pairwise disjoint, so the number of people is given by $S \geqslant 1 + |A_1| + |A_2| + \cdots + |A_n| = 3n + 1$. Thus, $3n + 1 \leqslant 16$, $n \leqslant 5$. The following chart shows that it is possible for $n = 5$:

Title number \ student	1	2	3	4	5	6	7	8	9	10	11	12	13	14	15	16
1	1	1	1	1	2	2	2	2	3	3	3	3	4	4	4	4
2	1	2	3	4	1	2	3	4	1	2	3	4	1	2	3	4
3	1	2	3	4	4	3	2	1	3	4	1	2	2	1	4	3
4	1	2	3	4	2	1	4	3	4	3	2	1	3	4	1	2
5	1	2	3	4	3	4	1	2	2	1	4	3	4	3	2	1

(5) Let the number of appearances of each $i (i = 1, 2, \ldots, 10)$ in $A_1, A_2, \ldots, A_k$ be $d(i)$. Obviously, we have

$$d(1) + \cdots + d(10) = |A_1| + |A_2| + \cdots + |A_k|$$
$$= 5 + 5 + \cdots + 5 = 5k,$$

so in order to know the range of k, we need to find the range of every $d(i)$.

Now we prove that $d(i) \leqslant 4 (i = 1, 2, \ldots, 10)$.

In fact, for $i \in M$, consider all the pairs (i, j) containing i (where $j \neq i$) and all the 5-element subsets containing i. On the one hand, for each i we have nine pairs (i, j) containing i; also every pair containing i will appear in $A_1, A_2, \ldots, A_k$ at most twice, so the total number of appearances of pairs (i, j) containing i is no more than $2 \times 9 = 18$.

On the other hand, among $A_1, A_2, \ldots, A_k$ there are $d(i)$ sets containing i; since $|A_j| = 5 (j = 1, 2, \ldots, k)$, each subset containing i will contain four pairs (i, j) containing i, and therefore $4 \cdot d(i) \leqslant 18$, thus $d(i) \leqslant 4$.

So $5k = d(1) + \cdots + d(10) \leqslant 4 \times 10 \Rightarrow k \leqslant 8$.

At last, when $k = 8$, the following eight 5-element number sets meet the requirement: $A_1 = \{1, 2, 3, 4, 5\}$, $A_2 = \{1, 6, 7, 8, 9\}$, $A_3 = \{1, 3, 5, 6, 8\}$, $A_4 = \{1, 2, 4, 7, 9\}$, $A_5 = \{2, 3, 6, 7, 10\}$, $A_6 = \{3, 4, 7, 8, 10\}$, $A_7 = \{4, 5, 8, 9, 10\}$, $A_8 = \{2, 5, 6, 9, 10\}$, so $k_{\min} = 8$.

Alternative solution: We consider the set-element relation table and connect any two $1s$ in the same column by a segment.

On the one hand, since each row has five $1s$, we get $C_5^2 = 10$ segments, so there are $10 \cdot k = 10k$ segments in this table. On the other hand, each segment corresponds to two rows where the end-points lie in; according to the assumptions, any two rows correspond to at most two segments, so the number of segments is no more than $2C_{10}^2 = 90$, thus $10k \leqslant 90$, hence $k \leqslant 9$.

If $k = 9$, every two rows happen to correspond to two segments, namely that each kind of segment happens to appear twice (here " *kind* " refers to the two rows a segment corresponds to). Therefore the $1s$ in row will appear in pairs, so the total number of $1s$ is even. Thus the number of $1s$ in the table is also even. But each column has five $1s$, so in the nine columns there are in total $9 \cdot 5 = 45$ $1s$, which is an odd number, which is a contradiction. Thus $k \leqslant 8$.

(6) (a) Suppose that airline $i (i = 1, 2, \ldots, k)$ offers a_i nonstop flights, and let $S = a_1 + a_2 + \cdots + a_k$. When airline i closes down, the total number of nonstop flights is $S - a_i$, so $S - a_i \geqslant n - 1$.

Therefore $\sum_{i=1}^{k} (S - a_i) \geqslant \sum_{i=1}^{k} (n-1) = k(n-1)$, so $kS - S \geqslant k(n-1)$, thus $S \geqslant \dfrac{k(n-1)}{k-1}$.

However S is integer, so $S \geqslant \left\lceil \dfrac{k(n-1) + k - 2}{k-1} \right\rceil = \left\lceil \dfrac{kn-2}{k-1} \right\rceil$.

On the other hand, we show that $S = \left\lceil \dfrac{kn-2}{k-1} \right\rceil$ meets the requirements, namely that there exists a graph G with order n, such that $\| G \| = \left\lceil \dfrac{kn-2}{k-1} \right\rceil$ (the number of edges), and that we can color the edges of G using k colors, such that after deleting the edges of any particular color (corresponding to airlines closing down), the graph still remains connected.

Induct on n with span $k - 1$.

When $n = 1, 2, \ldots, k-1$, we have $n < k$, so $S = \left\lceil \dfrac{kn-2}{k-1} \right\rceil = n + \left\lceil \dfrac{n-2}{k-1} \right\rceil = n$. Choose G to be an n-cycle and color its edges using k colors such that the colors of any two edges are different (this is possible because $n < k$). If we delete the edges any color, it is clear that the graph still remains connected. Supposing the conclusion is true when $n \leqslant r$ where $r \geqslant k - 1$, let us consider the situation where $n = r + k - 1$.

By induction hypothesis, we can choose a graph with vertices A_1, A_2, ... , A_r, which has $\dfrac{kn-2}{k-1}$ sides, such that we can color the edges using k colors and satisfy the requirements. Let the k colors be $1, 2, \ldots, k$.

Now we choose a vertex A_1 from G and add $k-1$ vertices B_1, B_2, ... , B_{k-1} aside from G, and connect the sides of $B_{i-1}B_i (i = 1, 2, \ldots, k$, where $B_0 = B_k = A_1)$, getting the graph G', so

$$\| G' \| = \| G \| + k = \left\lceil \frac{kn-2}{k-1} \right\rceil + k = \left\lceil \frac{k(n+k-1)-2}{k-1} \right\rceil.$$

Now if we color the edge $B_{i-1}B_i$ with the color i, we can verify that this coloring meets the requirements.

In fact, suppose that one deletes the edges with some color i ($i = 1, 2, \ldots, k$); consider any two peak numbers A and B in G'.

If $A, B \in V(G)$, then they are connected by induction hypothesis.

If $A, B \in \{B_1, B_2, \ldots, B_{k-1}\}$, since $B_0 = A_1$, and B_1, B_2, ... , B_{k-1} form a k-cycle whose edges have different colors, we know that only one edge is deleted in this circle, thus A and B are connected.

If $A \in V(G)$, $B \in \{B_1, B_2, \ldots, B_{k-1}\}$, then either $A = A_1$, or by induction hypothesis we know that A and A_1 are connected. Also B is connected with A_1 through the cycle $(A_1, B_1, B_2, \ldots, B_{k-1})$, so A and B are connected. Thus the conclusion holds when $n = r + k - 1$.

Therefore, these airlines offer at least $\dfrac{kn-2}{k-1}$ nonstop flights.

(b) When $n = 7$, $k = 5$, if for some airport there are only two nonstop flights connecting it with other airports, then when the two corresponding airlines close down, this airport will nor remain connected with others. Thus for each airport there are at least three nonstop flights, thus there are at least $3 \cdot 7 = 21$ nonstop flights. However, each nonstop flight is count twice, so $S \geqslant \dfrac{21}{2}$, but S is an integer, so $S \geqslant 11$.

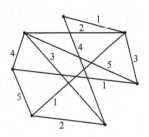

When $S = 11$, a possible construction is shown as in the graph, where $1 - 5$ represents five airlines.

In conclusion, there are at least 11 nonstop flights in total.

Exercise 10

(1) The score S of A is determined by the number of questions she answered correctly, so we can introduce some parameters. Suppose that the numbers of correct answers, unanswered questions and wrong answers are respectively x, y, z, where $x + y + z = 30$, then the score of A is $S = 5x + y + 0z = 5x + y$. The condition that "if the score of A is low but still exceeds 80, then B cannot infer the number of correctanswers A made" means that $S > 80$, that the corresponding x be the unique and that S is the minimum. Because S may correspond to (x, y) or $(x - 1, y + 5)$ or $(x + 1, y - 5)$, in order for (x, y) to be the unique possibility, we must have that the possibilities $(x - 1, y + 5)$ are $(x + 1, y - 5)$ nonexistent. As a result, we have $x - 1 < 0$ or $(x - 1) + (y + 5) > 30$. However $x \geqslant 1$, so we must have $(x - 1) + (y + 5) > 30$. Therefore, $x + y \geqslant 27$. Also since $(x + 1, y - 5)$ is nonexistent, we know $y - 5 < 0$ or $(x + 1) + (y - 5) > 30$. However, from $x + y + z = 30$ we can see that $x + y \leqslant 30$, thus we must have $y - 5 < 0$. Therefore $y \leqslant 4$, thus $S = 5x + y = 5(x + y) - 4y \geqslant 5 \times 27 - 4z \geqslant 5 \times 27 - 4 \times 4 = 119$. The equality holds when $x = 23$, $y = 4$. Thus $S_{min} = 119$, namely that in this exam A scored 119.

(2) We call $i + j$ the characteristic value of position (i, j); the position number of a student is then just the difference of the characteristic values between her former and latter positions. Now let M be the sum of characteristic values of all the positions, and let the initial empty position be (x, y), and the empty position after the adjustment be (p, q). We then have that the sum of all the characteristic values of students is $M - (x + y)$ before the adjustment and $M - (p + q)$ after the adjustment, thus $S = M - (x + y) - [M - (p + q)] = p +$

$q - (x + y)$. $S_{max} = 6 + 8 - (x + y)$, $S_{min} = 1 + 1 - (x + y)$, $S_{max} - S_{min} = [6 + 8 - (x + y)] - [1 + 1 - (x + y)] = 14 - 2 = 12$.

(3) Suppose that $A = \{1, 2, \ldots, 7\}$ is the set of possible show dates, and that A_i is the set of show dates for the ith troupe ($i = 1$, $2, \ldots, 12$), then we know that $A_1, A_2, \ldots, A_1$ do not contain each other. Otherwise suppose that A_i is contained in A_j, then the jth troupe cannot watch the show of the ith troupe. Consider the number of the shows as $t = \sum_{i=1}^{12} |A_i|$; if there is some $|A_i| = 0$, then the other troupes cannot watch the show of the ith troupe, which is a contradiction. Therefore, for any i, we have $|A_i| \geqslant 1$. Now we introduce parameters. Suppose that there are k sets A_i with one element, so that the number of the elements in each of the other sets is at least 2, then $t = \sum_{i=1}^{12} |A_i| \geqslant \underbrace{1 + 1 + \cdots + 1}_{}(k \text{ of } 1) + \underbrace{2 + 2 + \cdots + 2}_{12-k \text{ of } 2} = 24 - k$. We may as well suppose $\{a_1\} = A_1$, $\{a_2\} = A_2, \ldots, \{a_k\} = A_k$. Since $A_1, A_2, \ldots, A_{12}$ do not contain each other, $a_1, a_2, \ldots, a_k$ do not belong to $A_{k+1}, A_{k+2}, \ldots, A_{12}$, thus $A_{k+1}, A_{k+2}, \ldots, A_7$ are all the subsets of $\{a_{k+1}, a_{k+2}, \ldots, a_7\}$, namely that $\{A_{k+1}, A_{k+2}, \ldots, A_{12}\}$ is a family of subsets of $A \setminus \{a_1, a_2, \ldots, a_k\}$ which do not contain each other. Thus $C_{7-k}^{[(7-k)/2]} \geqslant 12 - k$, which is not true when $7 - k = 1, 2, 3, 4$, so $7 - k \geqslant 5$, namely $k \leqslant 2$, thus $t \geqslant 24 - k \geqslant 22$. Finally, when $t = 22$, we may let $A_1 = \{1\}$, $A_2 = \{2\}$, and choose A_3, $A_4, \ldots, A_{12}$, to be the 10 different 2-element sets of $\{3, 4, 5, 6, 7\}$, such as $A_3 = \{3, 4\}$, $A_4 = \{3, 5\}$, $A_5 = \{3, 6\}$, $A_6 = \{3, 7\}$, $A_7 = \{4, 5\}$, $A_8 = \{4, 6\}$, $A_9 = \{4, 7\}$, $A_{10} = \{5, 6\}$, $A_{11} = \{5, 7\}$, $A_{12} = \{6, 7\}$. Therefore, the minimum number of shows is 22.

(4) (i) Assume $n^2 = a_1^2 + a_2^2 + \cdots + a_k^2$, where $k = n^2 - 13$; we may as well suppose $a_1 \leqslant a_2 \leqslant a_3 \leqslant \cdots \leqslant a_k$. Then

$$a_k^2 = n^2 - (a_1^2 + a_2^2 + \cdots + a_{k-1}^2) \leqslant n^2 - (k - 1)$$
$$= n^2 - (n^2 - 14) = 14,$$

so $a_k \leqslant 3$. We may as well assume that in a_1, a_2, ..., a_k there are i 1s, j 2s and t 3s, where $i + j + t = k = n^2 - 13$, then

$$n^2 = (a_1^2 + a_2^2 + \cdots + a_k^2) = i + 4j + 9t = (n^2 - 13) + 3j + 8t,$$

so $3j + 8t = 13$, $8t = 13 - 3j \leqslant 13$, thus $t \leqslant 1$. When $t = 0$, $3j = 13$, which is a contradiction. When $t = 1$, $3j = 5$, which is a contradiction. As a result, for $k = n^2 - 13$, it is impossible to write n^2 as the sum of k positive squares, so $s(n) \leqslant n^2 - 14$.

(ii) If $s(n) = n^2 - 14$, then for any natural number $k \leqslant n^2 - 14$, the integer n^2 can be represented as the sum of k positive equates. Let us consider any positive integer $k \leqslant n^2 - 14$, let $k = n^2 - r(14 \leqslant r \leqslant n^2 - 1)$, we have to find an n and a partition $n^2 = a_1^2 + a_2^2 + \cdots + a_k^2$ (where $a_1 \leqslant a_2 \leqslant a_3 \leqslant \cdots \leqslant a_k$). Let us introduce parameters. We may set that in a_1, a_2, ..., a_k there are i 1s, j 2s and t 3s, where $i + j + t = k = n^2 - r$. We mark this k-partition as $k(i, j, t)$, then

$$n^2 = (a_1^2 + a_2^2 + \cdots + a_k^2) = i + 4j + 9t = (n^2 - r) + 3j + 8t;$$

therefore

$$3j + 8t = r. \qquad ①$$

One necessary condition that this requirement is met is that for all the $14 \leqslant r \leqslant n^2 - 1$, the equation ① has nonnegative integer solutions. In fact, if $r \equiv 1 \pmod 3$, let $r = 3r_1 + 1(r_1 \geqslant 5)$, then $(j, t) = (r_1 - 5, 1)$ is a solution of ①. If $r \equiv 2 \pmod 3$, let $r = 3r_1 + 2(r_1 \geqslant 4)$, then $(j, t) = (r_1 - 2, 1)$ is a solution of ①. If $r \equiv 0 \pmod 3$, let $r = 3r_1(r_1 \geqslant 5)$, then $(j, t) = (r_1, 0)$ is a solution of ①. We have noticed that $k = n^2 - r$, so ① is equal to

$$3j + 8t = n^2 - k. \qquad ②$$

Therefore, there is a partition $k(i, j, t)$ of n^2, thus j, t, k satisfy ②. Notice that there is no requirement for i in ②, so k, i, j, t also need to meet the requirement that $i = k - (j + t) \geqslant 0$, namely $k \geqslant j + t$. Therefore, for any n, only when the number k of terms is not less than $j + t$, where j, t meet the needs of ②, there can exist the corresponding

k to partition $k(i, j, t)$. We have noticed that $(n^2 - k) = 3j + 8t \geqslant 3(j + t)$, *thus one sufficient condition for* $k \geqslant j + t$ *is* $k \geqslant \dfrac{n^2 - k}{3}$, namely that $k \geqslant \dfrac{n^2}{4}$. So when $k \geqslant \dfrac{n^2}{4}$, for any n, there exists a corresponding partition $k(i, j, t)$. If the number of terms is $k > \dfrac{n^2}{4}$, then for n, the partition $k(i, j, t)$ does not necessarily exist, so we should find other forms of partitions. If $k < \dfrac{n^2}{4}$, then when n is small, there are fewer possible values of the number of terms. Therefore, we can examine them one by one beginning from smaller natural numbers n. In order to conveniently use Pythagorean numbers, we may take 3, 4, 5, 6, 8, 10, 12; but they are difficult to be decomposed. Let $n = 13$, then we get $13^2 = 13^2 = 12^2 + 5^2 = 12^2 + (3^2 + 4^2) = 8^2 + 8^2 + 5^2 + 4^2$. Also $8^2 = 4^2 + 4^2 + 4^2 + 4^2$, $4^2 = 2^2 + 2^2$, $2^2 = 1^2 + 1^2 + 1^2 + 1^2$, so 13^2 can be divided into 7, 10, 13, ..., 43 squares. Also $5^2 = 4^2 + 3^2$, so 13^2 can be divided into 5, 8, 11, ..., 44 squares. What is more, since $12^2 = 6^2 + 6^2 + 6^2 + 6^2$, $6^2 = 3^2 + 3^2 + 3^2 + 3^2$, $4^2 = 2^2 + 2^2 + 2^2 + 2^2$, $2^2 = 1^2 + 1^2 + 1^2 + 1^2$, we know that 13^2 can be divided into 3, 6, 9, ..., 33 squares. Finally, $13^2 = 3^2 + 3^2 + \cdots + 3^2 + 4^2$, and we choose 9 or 12 of all the 3^2 to be divided as $3^2 = 2^2 + 2^2 + 1^2$, thus 13^2 can be divided into $18 + 2 \times 9 = 36$, $18 + 2 \times 12 = 42$ squares. Therefore, 13^2 can be divided into 1, 2, ..., 44 squares. However for $k \geqslant 45$, we have $k \geqslant \dfrac{13^2}{4}$, so 13^2 can be divided into $k(i, j, t)$. Thus $s(13) = 13^2 - 14$.

(iii) We prove that when $n = 2^m \times 13$, $s(n) = n^2 - 14$. In fact,
$$n^2 = (2^m \times 13)^2 = 4^t (2^{m-t} \times 13)^2 (0 \leqslant t \leqslant m).$$

Since 13^2 can be divided into 1, 2, 3, ..., 155 squares, we know that n^2 can be divided into 1, 2, 3, ..., $4^m \times 155$ squares. But $4^m \times 155 > \dfrac{(2^m \times 13)^2}{4} = \dfrac{n^2}{4}$, from the previous discussion we know that n^2 can be divided into 1, 2, 3, ..., $n^2 - 14$ squares, so $s(2^m \times 13) = (2^m \times 13)^2 - 14$.

(5) When $n = 4$, we can choose $a_1 = a_2 = a_3 = a_4 = 2$, then

$$a_1 + a_2 + a_3 + a_4 = 8 \geqslant 4, \; a_1^2 + a_2^2 + a_3^2 + a_n^2 = 16 \geqslant 4^2.$$

So a_1, a_2, a_3, a_4 meet the conditions of the problem, and max $\{a_1, a_2, \ldots, a_n\} = 2$. Next let us show that for any real number a_1, $a_2, \ldots, a_n (n > 3)$, that satisfy $a_1 + a_2 + \cdots + a_n \geqslant n$ and $a_1^2 + a_2^2 + \cdots + a_n^2 \geqslant n^2$ we have max $\{a_1, a_2, \ldots, a_n\} \geqslant 2$. Suppose max $\{a_1, a_2, \ldots, a_n\} < 2$, and there are i nonnegative numbers in a_1, $a_2, \ldots, a_n$, denoted by $x_1, x_2, \ldots, x_i$, and there are j negative numbers, denoted by $-y_1, -y_2, \ldots, -y_j$, where $y_1, y_2, \ldots, y_j > 0$, $i \geqslant 0$, $j \geqslant 0$, $i + j = n$. Thus max $\{x_1, x_2, \ldots, x_i\} < 2$. Since

$$x_1 + x_2 + \cdots + x_i + [(-y_1) + (-y_2) + \cdots + (-y_j)] \geqslant n,$$

we have $x_1 + x_2 + \cdots + x_i \geqslant n + y_1 + y_2 + \cdots + y_j$. Also max $\{x_1, x_2, \ldots, x_i\} < 2$, $y_1, y_2, \ldots, y_j > 0$, therefore

$$2i = 2 + 2 + \cdots + 2 > x_1 + x_2 + \cdots + x_i \geqslant n + y_1 + y_2 + \cdots + y_j$$
$$= i + j + y_1 + y_2 + \cdots + y_j.$$

This implies $i - j > y_1 + y_2 + \cdots + y_j$. Since

$$x_1^2 + x_2^2 + \cdots + x_i^2 + (-y_1)^2 + (-y_2)^2 + \cdots + (-y_j)^2 \geqslant n^2,$$

we have

$$x_1^2 + x_2^2 + \cdots + x_i^2 \geqslant n^2 - (y_1^2 + y_2^2 + \cdots + y_j^2)$$
$$\geqslant n^2 - (y_1 + y_2 + \cdots + y_j)^2$$
$$> n^2 - (i - j)^2$$
$$= (i + j)^2 - (i - j)^2 = 4ij.$$

Also $i \geqslant 0$, therefore $j < 1$, so $j = 0$, thus $a_1, a_2, \ldots, a_n$ are all nonnegative numbers. Thus $0 \leqslant a_i < 2 (i = 1, 2, \ldots, n)$, thus $4n > a_1^2 + a_2^2 + \cdots + a_n^2 \geqslant n^2$, we get that $n < 4$, which contradicts the condition $n > 3$. In conclusion, max $\{a_1, a_2, \ldots, a_n\} = 2$.

Exercise 11

(1) Suppose that the number of male students in the ith row is a_i, then the number of female students is $75 - a_i$. According to the assumption we know that

$$\sum_{i=1}^{22} (C_{a_i}^2 + C_{75-a_i}^2) \leqslant 11 \times C_{75}^2,$$

because for any two columns the number of pairs of students in these two columns and the same column having the same gender is no more than 11. Thus we get

$$\sum_{i=1}^{22} (a_i^2 - 75a_i) \leqslant -30525,$$

thus $\sum_{i=1}^{22} (2a_i - 75)^2 \leqslant 1650$. By the Cauchy-Schwartz we know that

$$\left[\sum_{i=1}^{22} (2a_i - 75)\right]^2 \leqslant 22 \sum_{i=1}^{22} (2a_i - 75)^2 \leqslant 36300,$$

therefore $\sum_{i=1}^{22} (2a_i - 75) < 191$, hence $\sum_{i=1}^{22} a_i < \dfrac{191 + 1650}{2} < 921$.

So the number of boy students is less than 928.

(2) We use vertices to represent people, and for any two people who have talked to each other, we connect them by an edge, then obtaining a simple graph. Now we calculate the number S of the angles in this graph. On the one hand, since each vertex is related to $3k + 6$ sides, thus we have C_{3k+6}^2 angles. Since there are $12k$ vertexes, the total number of angles will be $12k C_{3k+6}^2$. Now since every angle has a vertex and different vertices correspond to different angles, we have $S = 12k C_{3k+6}^2$. On the other hand, since for any two people, the numbers of people who talked to them are the same, say t, then each two-people group correspond to t angles. Also each angle corresponds to only one two-people group and different two-people groups correspond to different angles, so $S = t C_{12k}^2$, thus

$$12k\,C_{3k+6}^2 = t\,C_{12k}^2, \ t = \frac{(3k+6)(3k+5)}{12k-1},$$

thus

$$16t = \frac{(12k-1+25)(12k-1+21)}{12k-1}$$

$$= (12k-1) + 25 + 21 + \frac{25 \times 21}{12k-1}.$$

Obviously, $(3, 12k-1) = 1$, so $12k-1 \mid 25 \times 7$. We have noticed that $12k-1$ is congruent to 3 mod 4, thus $12k-1 = 7, 5 \times 7, 5^2 \times 7$, among which only $12k-1 = 5 \times 7$ has integer solutions $k = 3$, $t = 6$, so the possible number of people is 36. Now we prove that having 36 people is possible, namely there exists a graph G of order 36, such that the degree of each vertex is 15 and each pair of vertices corresponds to exactly six angles. First we construct six complete graphs K_6, and use the number 1, 2, 3, 4, 5, 6 to denote the vertices of each graph. Now connect some vertices of the six graphs, forming a graph G. The vertices of it are marked as (i, j), which represents the jth vertex in the ith graph. We notice that each vertex is connected with the five vertices with the same x-coordinate; now we connect each vertex with the five vertices sharing the same y-coordinate, and the five vertices with the same coordinate difference (namely the same $i-j$ value). Therefore, each vertex has 15 edges, and each vertex has precisely two edges connecting it to each complete graph not containing it. For any two vertices, if they are both connected with some third vertex, then we call that these three vertices form an angle. Now let us prove that any two vertices in G correspond to six angles. In fact, inspecting any two vertices (i, j), (i', j'), if $i = i'$, then they are in the same diagram and therefore they correspond to four angles in this diagram. In addition, there are two vertices $(i'-j'+j, j)$ and $(i-j+j', j')$ that are connected with both of them.

 If $i \neq i'$ and $i-j = i'-j'$, then (i, j') and (i', j) are connected with both points; also for any i'' not equaling i or i', the vertex $(i''$,

$i'' - i + j$) is connected with both points, so they also correspond to six angles in total. If $i \neq i'$, and $i - j \neq i' - j'$, then (i, j'), (i', j), $(i, i' - i' + j')$ and $(i', i' - i + j)$ are connected with both points; also there are two vertices $(i' - j' + j, j)$, $(i - j + j', j')$ connected with both and that they are the only such vertices, so they also correspond to six angles. In conclusion, the number of people attending the meeting is 36.

(3) From the condition that the total score of each player is 26, we may calculate the sum S of the total scores of the n players after the kth day. On the one hand, the scores in each day are $1 + 2 + 3 + \cdots + n$, so $S = k(1 + 2 + \cdots + n)$. On the other hand, since each player has s total score of 26, we know $S = 26n$, thus, $k(n + 1) = 52$, so $(n, k) = (51, 1)$, $(25, 2)$, $(12, 4)$, $(3, 13)$. When $(n, k) = (51, 1)$, the total score of each competitor is different from each other, which is impossible. From the following constructions we know that the other three situations are possible. Namely, when $(n, k) = (25, 2)$, the score set of the ith competitor can be $A_i = \{i, 26 - i\}(i = 1, 2, \ldots, 25)$; when $(n, k) = (12, 4)$, the score set of the ith competitor can be $A_i = \{i, 13 - i, i, 13 - i\}(i = 1, 2, \ldots, 12)$ and when $(n, k) = (3, 13)$, we can have $A_1 = \{2, 3, 1\} \cup \{1, 3, 1, 3, \ldots, 1, 3\}$, $A_2 = \{3, 1, 2\} \cup \{2, 2, 2, 2, \ldots, 2, 2\}$, $A_3 = \{1, 2, 3\} \cup \{3, 1, 3, 1, \ldots, 3, 1\}$.

(4) Suppose that there are t senior students and each student has k friends. For convenience, we use 30 vertices to represent the 30 students. For any two vertices A and B, if they are friends and A is older than B, we connect them by a directed edge from A to B, thus obtaining a directed graph. Call the vertices corresponding to senior students big vertices. Suppose that all the big vertices are $A_1, A_2, \ldots, A_t$, from the assumption we can get

$$d^+(A_i) > d^-(A_i), \quad d(A_i) = d^+(A_i) + d^-(A_i) = k,$$

so $d^+(A_i) \geqslant \dfrac{k+1}{2}$. We may as well suppose that the age of student A_i is increasing with I, then $d^+(A_1) \leqslant 30 - t$ (the number of the outgoing edges from A_1 is at most $30 - t$ since they must not end with $A_1, A_2, \ldots, A_t$), and $d^+(A_t) = k$. We will compute the sum S of the outgoing degrees of all the big vertices. On the one hand

$$S = d^+(A_1) + d^+(A_2) + \cdots + d^+(A_{t-1}) + k \geqslant \frac{k+1}{2}(t-1) + k.$$

On the other hand, $S \leqslant \|G\| = 15k$, so $15k = \|G\| \geqslant S \geqslant \dfrac{k+1}{2}(t-1) + k$ hence

$$t \leqslant \frac{28k}{k+1} + 1. \qquad\qquad ①$$

In addition, $\dfrac{k+1}{2} \leqslant d^+(A_1) \leqslant 30 - t$, so

$$k \leqslant 59 - 2t. \qquad\qquad ②$$

From ① and ② we can eliminate k (by monotonicity with k of the function on the right of ①), getting $t \leqslant \dfrac{28(59-2t)}{60-2t} + 1$, and that

$$t^2 - 59t + 856 \geqslant 0. \qquad\qquad ③$$

However $t \leqslant 30$, and the maximum integer making ③ right is $t = 25$, so the number of older students is not more than 25; finally it is possible for $t = 25$. Actually when $t = 25$, the above inequality is correct, and when we put ② in it, we can get the solution $k = 9$.

Arrange $1, 2, \ldots, 30$ into six rows (see the diagram),

1,	2,	3,	4,	5
6,	7,	8,	9,	10
11,	12,	13,	14,	15
16,	17,	18,	19,	20
21,	22,	23,	24,	25
26,	27,	28,	29,	30

Let i and j be one pair of friends if and only if i and j satisfy one of the following three conditions:

(i) i and j are in the adjacent rows but not in the same row column;

(ii) i and j are in the same column, and one of them is in the last row;

(iii) i and j are both in the first row.

Now each person has nine friends; for example, the nine friends of 1 are 2, 3, 4, 5, 7, 8, 9, 10 and 26, so the maximum of t is 25.

(5) Suppose that there are n participants. The answer sheet of each participant is a sequence with length 4, which is composed of the three letters A, B and C. As a result, the question is equivalent to coloring the cells of an $n \times 4$ check board, so that for any three rows, there are three cells in these three rows that lie in the same column and have different colors. Suppose that the total number of pairs that lie in the same column and have different colors is S. On the one hand, for any one column, suppose there are a cells of color A, b cells in color B and c cells in color C, then the number of pairs with different colors is $ab + bc + ca$. We notice that

$$3(ab + bc + ca) \leqslant a^2 + b^2 + c^2 + 2(ab + bc + ca)$$
$$= (a + b + c)^2 = n^2,$$

so there are at most $\dfrac{4n^2}{3}$ pairs of different colors in the four rows, namely that $S \leqslant \dfrac{4n^2}{3}$. On the other hand, for any three rows, there is one column containing three colors, which form three pairs, so the total number of pairs is $3C_n^3$. However, the same pair may appear in $n - 2$ different 3-row groups, thus $S \geqslant 3 \cdot \dfrac{C_n^2}{n - 2}$, so $3 \cdot \dfrac{C_n^2}{n - 2} \leqslant S \leqslant \dfrac{4n^2}{3}$. But this inequality is always true, so we cannot get the range of n from it. Readers may consider whether this estimate can be improved. Considering the problem from another perspective, we may try to find, by contradiction, three rows such that for all columns, the intersections of the three rows with this column contain at most two colors. After our attempt, we find that $n < 10$. In fact, assume the contrary. Choose any 10 rows to get a diagram M with 10×4 cells. Considering the first column in the diagram M, we know that at least one color appears at most three times, so there are at least seven rows, such that first cells only have two colors. Considering the second cells of the seven rows, there must be at least one color that appears at most twice, therefore in these seven rows there are at least five rows whose second cells only have two colors. Next consider the third line of the five rows, there must be at least one color that appears at most once, so in these five rows, there are at least four rows whose third cells only have two colors. Finally consider the last cells of the four rows, there must be at least one color that appears at most once. Therefore in these four rows there are at least three rows such that for each column, the three cells at the intersections of these three rows and the column have only two colors, which is a contradiction. When $n = 9$, the answers to each question chosen by the 9 people can be respectively $(1, 2, 1, 2)$, $(2, 3, 2, 2)$, $(3, 1, 3, 2)$, $(1, 1, 2, 1)$, $(2, 2, 3, 1)$, $(3, 3, 1, 1)$, $(1, 3, 3, 3)$, $(2, 1, 1, 3)$, $(3, 2, 2, 3)$, so the maximum value of n is 9.

(6) Let us interpret the condition "*In every two committee members, there is at most one public member*" as $|A_i \cap B_j| \leqslant 1$. We can interpret A_i as the set of members of the ith committee, and $F = \{A_1, A_2, \ldots, A_k\}$. Let X be the set of the 25 people. Let us calculate the total number S of all the two-people groups. On the one hand, $|A_i| = 5$, so there are $C_5^2 = 10$ two-people groups from each of A_i. Therefore, since in F there are k sets, we have $10k$ two-people groups. Because $|A_i \cap B_j| \leqslant 1$, $10k$ two-people groups must be different from each other, so $S \geqslant 10k$. On the other hand, since $|X| = 25$, we know that the total number of the two-people groups in X is $S = C_{25}^2$, thus $C_{25}^2 = S \geqslant 10k$, so $k \leqslant 30$, this completes the proof.

Exercise 12

(1) (i) If $k \geqslant 7$, there are $2^k - 1$ nonempty subsets in A, among which the biggest subset sum is equal to or less than $17k$, but $2^k - 1 > 17k$, so there must exist two subsets whose sums are the same, which is a contradiction. If $k = 6$, consider the subsets of A with at most four elements of A. There are $C_6^1 + C_6^2 + C_6^3 + C_6^4 = 56$ different subsets, whose sum of elements lie in the internal $[1, 57]$, because any of such sum is $\leqslant 16 + 15 + 14 + 13 = 58$, and that from $13 + 16 = 15 + 14$ we know that 13, 14, 15 and 16 do not all belong to A. If $1 \in A$, from $1 + 15 = 16$ we know that 15 and 16 do not both belong to A; from $1 + 13 = 14$ we know that 13 and 14 do not both belong to A, and from $1 + 11 = 12$ we know that 11 and 12 do not both belong to A. Thus the maximum sum is no more than $16 + 14 + 12 + 10 = 52$, but $56 > 52$, so there must be two subsets whose sums are the same, which is a contradiction. If $2 \in A$, from $2 + 14 = 16$, 14 and 16 do not both belong to A; from $2 + 13 = 15$, 13 and 15 do not both belong to A, and from $2 + 10 = 12$, 10 and 12 do not both belong to A. Thus the maximum sum is no more than $16 + 15 + 12 + 9 = 52$, but $56 > 52$, so there must be two subsets whose sums are the same, which is a contradiction. If neither 1 nor 2 belongs to A, then the minimum sum is not less than 3, therefore the sums are all in the internal $[3, 57]$, so there are at most 55 different sums, but $56 > 55$, so there must be two subsets with the same sum, which is a contradiction. Therefore $k \leqslant 5$.

(ii) Suppose that the sum of A is S. If $S < 16$, consider $B = A \cup \{16\}$, which is a $(k + 1)$-element subset containing A. Since the sums of any two subsets in A are different, and that the sum of any subset of B containing 16 is bigger than that of any subset not containing 16, we know that any two subsets in B does not have the same sum, which is a contradiction. Thus $S \geqslant 16$. Since $A = \{1, 2, 4, 9\}$ meets the condition, and $S(A) = 16$, we know that the minimum of S is 16. If $k \leqslant 4$, then $S \leqslant 16 + 15 + 14 + 13 = 58 < 66$; if $k = 5$, and that 16 and 15 do not both belong to A, then $S \leqslant 16 + 14 + 13 + 12 + 11 = 66$; if $k = 5$, and that 16 and 15 both belong to A, then any two number in $(14, 13)$, $(12, 11)$, $(10, 9)$ do not both belong to A, thus $S \leqslant 16 + 15 + 14 + 12 + 10 = 67$. Equality cannot here, otherwise 14, 12, 10, $16 \in A$, but $16 + 10 + 12 + 14$, which is a contradiction. So $S \leqslant 66$. Since $A = \{16, 15, 14, 12, 9\}$ satisfies the conditions, and $S(A) = 66$, we know that the maximum of S is 66.

(2) The various amounts proposed by each councilor form a sequence with length 200, and the sequences filled by the 2000 councilors form a 2000×200 table, in which the sum of each row is not more than S. Now we add another row at the bottom of the table, such that each number in this new row is less than or equal to at least k existing numbers in the same column. We want to find the minimum k such that the sum of numbers in the new row does not exceed S. We call such k a good number. Obviously $k = 2000$ is good; moreover, $k = 1999$ is also good. In fact when $k = 1999$, there is only one number in each column which is larger than x_j (the corresponding number in the new row), and we call these numbers bad numbers. Therefore, there are at most 200 bad numbers from 200 columns. However, since there are 2000 rows, thus there is at least one row without bad numbers, which indicates that the councilor who proposes this row agrees to the each chosen amount. In the same way (that is, we gradually narrow the encirclement), we will find that $k = 1991$ is good. In fact, when $k = 1991$, for each column there are at most nine bad numbers and thus there are at most 1800 bad number from 200 columns, so there is at least one row without bad numbers. Next we prove that $k = 1990$ is not good. In fact, let us combine 10 rows of the table into one group, and in each row of the ith group, we fill the numbers $\frac{199}{S}, \frac{199}{S}, \ldots, \frac{199}{S}$, $0, \frac{199}{S}, \ldots, \frac{199}{S}$, where only the number in the ith column is 0, and other numbers are all $\frac{199}{S}$. In this numerical table, each row has

199 $\dfrac{199}{S}s$ and one 0, so the sum of each row is S. In each column there

are 10 0s and 1990 $\dfrac{199}{S}s$, so for any j, we can choose $x_j = \dfrac{199}{S}$. Now

the approved total amount of expenditure is $200 \cdot \dfrac{199}{S} > S$, which is a

contradiction. All in all, the minimum of k is 1991.

(3) If we disqualify all the other teams except team i, then
obviously team i is the champion. Therefore $f_i \leqslant n-1$, and $F \leqslant n(n-1)$.
When all matches end up in draws, for any $i = 1, 2, \ldots, n$, we have
$f_i = n-1$, so $F = n(n-1)$. Thus $F_{max} = n(n-1)$. When $n \geqslant 5$, firstly
prove $F \geqslant n$. One the sufficient condition for $F \geqslant n$ is $f_i \geqslant 1$, namely
that there is no team whose score is strictly more than that of others.
In addition, suppose that for $i = 1, 2, \ldots, n-1$, the score S_n of the
nth team is more than the score S_i of the ith team, then in order for i
team to be the champion, we have to disqualify at least one team, thus
$f_i \geqslant 1 (i < n)$ and $f_n = 0$, so $F \geqslant n-1$. If $F = n-1$, then equality
must hold, namely that when $i = 1, 2, \ldots, n-1$, $f_i = 1$, and $f_n = 0$.
Suppose that the team we need to disqualify in order that team
$i (i < n)$ becomes the champion is a_i, then we have the following:

(i) $S_i \geqslant S_n - 2$. In fact, if $S_n > S_i + 2$, after disqualifying team
a_i, the score of team n is at least $S_n - 3$, while the score of team
i remains the same, but $S_n - 3 \geqslant S_i$, which contradicts the assumption
that team i becomes the champion.

(ii) When $a_i \neq n$, team a_i is defeated by team n. Otherwise, after
disqualifying a_i, the score of team n is at least $S_n - 1$, while the score
of team i remains the same, but $S_n > S_i$, so $S_n - 1 \geqslant S_i$, which
contradicts the fact that team i becomes the champion.

(iii) When $i \neq j$, $a_i \neq a_j$. Otherwise, after disqualifying a_i, there
are two champion teams, (team i and team j), which is a contradiction.

From (ii) and (iii) we know that, among $a_1, a_2, \ldots, a_{n-1}$, the
teams which are not team n are all defeated by team n, so there are at

least $n-2$ teams defeated by team n, so $S_n \geqslant 3(n-2)$. Therefore,

$$S = \sum_{i=1}^{n} S_i = \sum_{i=1}^{n-1} S_i + S_n \geqslant \sum_{i=1}^{n-1} (S_n - 2) + S_n$$

$$= (n-1)(S_n - 2) + S_n$$

$$= nS_n - 2(n-1) \geqslant n \cdot 3(n-2) - 2(n-1)$$

$$= 3n^2 - 8n + 2.$$

On the other hand, the contribution to S of each match is at most 3, so $S \leqslant 3C_n^2 = 3\dfrac{n(n-1)}{2}$, thus $3\dfrac{n(n-1)}{2} \geqslant 3n^2 - 8n + 2$, so $n < 5$, which is a contradiction.

Finally, when A_1 wins A_2, A_2 wins A_3, ..., A_n wins A_1, and the rest of matches end up in draws, we can prove that all $f_i = 1$, thus $F = n$, so when $n \geqslant 5 F_{\min} = n$.

Exercise 13

(1) The minimum of n is 25. Denote the four choices of each question by 1, 2, 3, 4, and denote the answers in each answer sheet by (g, h, i, j, k), where $g, h, i, j, k \in \{1, 2, 3, 4\}$. For all the 2000 answers (g, h, i, j, k), we collect all those whose last four components are the same into one category; there are thus $4^4 = 256$ categories in total. Since $2000 = 256 \times 7 + 208$, there must be eight answer sheets belonging to the same category A. We take out the eight answer sheets; in the remaining 1992 ones, there are still 8 belonging to the same category B. Taking these eight answer sheets again, in the remaining 1984 ones, there are still 8 ones belonging to the same category C. These 24 answer sheets belong to three categories, A, B and C. Now when $n \leqslant 24$, we take any n answer sheets from these 24 of them, then for any 4 of the n answer sheets, two of them must be in the same category, so the requirement is not satisfied. Thus $n \geqslant 25$. Next let us construct 2000 answer sheets, such that they have 250 different answers in total, and that the number of answer sheets with any answer is 8. Here the

250 answers are chosen from the $4^4 = 256$ different answers (g, h, i, j, k) which satisfies $g + h + i + j + k \equiv 0 \pmod 4$. Obviously, for any two different answers, they have at most three same components. Otherwise, there are four same components, thus from the congruence property we know that the fifth component is also the same, which is a contradiction. If we choose any 25 from the 2000 answer sheets, since there are at most 8 of them having the same answer, we know that at least four answer sheets are pairwise different, so for any two of the four, there are three common answers. Therefore the requirement is satisfied when $n = 25$.

(2) The sequence n, n, $n-1$, $n-1$, ..., 2, 2, 1, 1, 2, 2, ..., $n-1$, $n-1$, n, n meets the requirement, since for any $a_i = a_j$, it must be that i and j are either adjacent or symmetric with respect to the 1, 1 in the middle. Therefore the maximum of k is not less than $4n - 2$. Now note that, in any sequence that meets the requirement, any three consecutive terms cannot be the same. Also, if all the adjacent terms of a particular term (note that both the first term and last term have only one adjacent term) are different from it, we may add a same term besides this term, so that the new sequence still meets the requirement. Therefore, we only need to consider the sequences such that each term is repeated immediately before or after it. Furthermore, by removing one from two consecutive and identical terms, we just need to consider the sequences such that any two consecutive terms are different. Now prove by induction that: if any two consecutive terms are different in a sequence a_1, a_2, ..., a_k, each term is a natural number equal to or less than n, and there do not exist subscripts $p < q < r < s$, such that $a_p = a_r \neq a_q = a_s$. Then $k \leqslant 2n - 1$. When $n = 2$, the conclusion is obviously true. Suppose that the conclusion is true for all natural numbers less than n, consider the situation of n. Suppose that u_1, a_2, ..., a_k is a longest sequence which meets the requirement, $a_i \in \{1, 2, ..., n\}$ $(i = 1, 2, ..., k)$. Let $a_k = t$, $1 \leqslant t \leqslant n$. In $a_1, a_2, ...$, a_{k-1}, if any of them does not equal t, then we can add one t before the

term a_1, getting a much longer sequence, which is a contradiction. Therefore, in a_1, a_2, $\ldots$, a_{k-1}, there is at least one term that equals t. Suppose a_v is such a term whose subscript is the biggest then none of a_{v+1}, a_{v+2}, $\ldots$, a_{k-1} is t. If $v = 1$, then none of a_2, a_3, $\ldots$, a_{k-1} equals t, so by induction hypothesis we know that $k - 2 \leqslant 2m - 1$, so $k \leqslant 2(m + 1) - 1$, thus the conclusion is true for n. If $v > 1$, then let $A = \{a_1, a_2, \ldots, a_v\}$, $B = \{a_{v+1}, a_{v+2}, \ldots, a_{k-1}\}$, so that in both A and B, there are no two consecutive items which are identical, and that all terms of B are in $\{1, 2, \ldots, n\}\backslash\{t\}$, and all terms of A are in $\{1, 2, \ldots, n\}\backslash\{B\}$, therefore we can use inductive hypothesis. Suppose that for A, the number of different terms is p, and that for B the number is q, then $p + q \leqslant n$, and that $v \leqslant 2p - 1$, $k - v - 1 \leqslant 2q - 1$. So $k \leqslant 2q + v \leqslant 2q + 2p - 1 = (p + q) - 1 \leqslant 2n - 1$. As a result, the maximum k is $4n - 2$.

(3) A sequence can be regarded as a permutation. For two different permutations, one sufficient condition for any of them not being a section of the other is that they have the same length. Suppose that their lengths are both r, then there are 2^r different permutations. However, the number of sequences given in the problem is 2^n, so $r = n$, namely that there are 2^n different permutations whose lengths are n. Among them any one is not a section of any other, and $S = n \times 2^n$. Now we prove $S \geqslant n \times 2^n$. We need to perform a sequence of operations so that after the operations. The length of each permutation is not less than n. We call a permutation whose length is n a standard one, one whose length is not n a short one or a long one. If we have short permutations, then we must also have long permutations. Otherwise, each short permutation can be extended to at least two different standard permutations, making the number of standard permutations exceed 2^n, which is a contradiction. Now take out any short permutation A, there must exist a long one B. Now we shall replace A and B, and add two new permutations $A \cup \{0\}$, $A \cup \{1\}$, so that the requirement is still satisfied. Because $|A| < n$, $|B| > n$, we know that the

increment of the length sum S after our operation is $|A|+1+|A|+1-|A|-|B|=2+|A|-|B|\leqslant 2+(n-1)-(n+1)=0$, thus S does not increase. We may repeat this till there is no short permutation, so we have $f \geqslant f' \geqslant n \times 2^n$.

(4) $r(m, n) = m + n - 2$. On the one hand, we know that it is possible for $r = m + n - 2$ (just put a stone at each position of the first row and the first column except for their intersection). Now let us prove $r < m + n - 1$, namely that when there are at least $m + n - 1$ stones on the check board, at least one chess can be removed. We induct on $m + n$. When $m = 2$, $n = 2$, there are at least three stones on the board, and obviously there is a removable stone. Suppose that the conclusion holds for the integers (m, n) such that $m + n < k$. Consider the situation $m + n = k$. Now, there are at least $k - 1$ stones on the board. If $n = 2$ or $m = 2$, then the conclusion is obviously true. If $m > 2$ and $n > 2$, then since there are at least $m + n - 1 > m$ stones on the board, there must be one row containing at least two stones. We may as well assume that the first row has the most stones that there are t stones in this row, and that they are denoted by $a_{11}, a_{12}, \ldots, a_{1t}$ ($t > 1$). We know that there is no other stone in the columns where these stones lie in; otherwise assume that there is one more stone in the column where a_{11} lies in, then a_{11} can be removed. Therefore, we may remove the first row and first t column, leaving the chessboard of size $(m - 1) \times (n - t)$, on which there are at least $m + n - 1 - t = (m - 1) + (n - t)$ stones. We notice that $(m - 1) + (n - t) = k - t - 1 < k$. If $n - t \geqslant 2$, we may apply the induction hypothesis to $(m - 1) \times (n - t)$, so that we know that there exists a stone in the $(m - 1) \times (n - t)$ chessboard that can be removed. If $n - t \leqslant 1$, then all the stones are distributed in the first row or the nth column, from which we can also get that there is stone that can be removed, hence our proposition gets proved.

(5) Because $|A_i \cap B_j| \leqslant 1$, let us calculate the number S of our

binary subsets (or 2-element subsets). On the one hand, $|A_i| = 3$, so in each A_i there are $C_3^2 = 3$ binary subsets, so we have a total of $3k$ binary subsets. Because $|A_i \cap B_j| \leqslant 1$, these $3k$ binary subsets are different from each other. Hence $S \geqslant 3k$. On the other hand, suppose $|X| = n$, then in X the total number of the binary subsets is $S = C_n^2$.

So $C_n^2 = S \geqslant 3k$, thus $k \leqslant \dfrac{n^2 - n}{6}$. Next we need to construct an F

which meets the conditions, such that $|F| \geqslant \dfrac{n^2 - 4n}{6}$. In order to

construct F, what we need to do is to construct several 3-number groups $\{a, b, c\}$, such that $|A_i \cap A_j| \leqslant 1$. Let us consider the opposite case when $|A_i \cap A_j| \geqslant 2$. The third elements in A_i and A_j shall be different from each other, otherwise $A_i = A_j$. Thus, the idea is to make sure that this does not happen, namely that when A_i and A_j have two common elements, their third elements must also be equal. This suggests that we only have two free variables in a, b and c, and that the third component c is uniquely determined by the first two components a and b. From this we can imagine that a, b, c may satisfy some equation. The simplest equation is a linear: $a + b + c = 0$. However, a, b, and c are all positive integers, so this equation cannot hold. Now we can change this equation into a congruence relation: $a + b + c \equiv 0 \pmod{n}$. Let us inspect all the 3-number groups $\{a, b, c\}$, satisfying $a + b + c \equiv 0 \pmod{n}$. Each such 3-number group constitutes one set, so that these sets meet the requirement. Now let us prove that

the number of such sets is no less than $\dfrac{n^2 - 4n}{6}$. We notice that this set

is uniquely decided by two numbers such as a and b, from the 3-number group $\{a, b, c\}$, which satisfies $a + b + c \equiv 0 \pmod{n}$, which is because

$$c = \begin{cases} n - (a + b) & (a + b < n), \\ 2n - (a + b) & (a + b \geqslant n). \end{cases} \qquad \textcircled{1}$$

Therefore, we just need to consider the possible values that a and b can take. Firstly, there are n values for a. After a is fixed, then the value of b shall meet the following two conditions: (i) $b \neq a$; (ii) the value of b shall be such that $c \neq a$ or b, where c is decided by ①. Obviously, it is sufficient for (ii) that $n - (a + b) \neq a$ or b, and $2n - (a + b) \neq a$ and b, namely that $b \neq n - 2a$ or $\dfrac{n-a}{2}$, and $b \neq 2n - 2a$ or $\dfrac{2n-a}{2}$. But $n - 2a$ and $2n - 2a$ do not both belong to X, since otherwise $1 \leqslant n - 2a \leqslant n$, $1 \leqslant 2n - 2a \leqslant n$, thus $2a + 1 \leqslant n \leqslant 2a$, which is a contradiction. As a result, for there are at most three values of b that do not meet the requirement (ii). Together with (i), for b there are at most four values that cannot be chosen, so there are at least $n - 4$ values. Thus, there are at least $n(n - 4)$ such 3-number groups. However, each 3-number group $\{a, b, c\}$ is counted at most six times, so there are at least $\dfrac{n(n-4)}{6}$ 3-number groups which satisfy the conditions. All in all, the proposition gets proved.

(6) After we finish filling the first row and the first column of the data table, the table will be uniquely determined. Thus when $r = m + n - 1$, there exists a corresponding way of filling. Next we prove that for any way of filling which satisfy the conditions, we must have $r \geqslant m + n - 1$. Assume the contrary, we will prove that when $r \leqslant m + n - 2$, no matter how one fills r cells in the table, the rest of the table is not uniquely determined. Induct on $m + n$. When $m + n = 4$, $m = n = 2$, in the 2×2 table we fill in two numbers, thus the table is not uniquely determined. Suppose that the conclusion is true for $m + n = k$. Now consider the situation where $m + n = k + 1 > 4$. In the table, we fill in $r \leqslant m + n - 2 = k - 1$ numbers, and we want to prove that this table is not uniquely determined. In order to use the induction hypothesis, we shall delete one row or one column from the table, and that this row or column shall have the following properties:

(i) After deleting this row, there must be at least two rows in this table (otherwise we cannot use the induction hypothesis). Here we may as well suppose $m \leqslant n$, thus $n > 2$, so we can delete one column.

(ii) After we delete this column, in the remaining table of size $(m-1) \times n$, there are no more than $(m-1) + n - 1 = k - 2$ numbers, namely that there is at least one number in the deleted column.

(iii) When we add back this column, the table is still not uniquely determined, which requires that there are at most one number in this column (this is a sufficient condition). In fact, suppose there is a number a which is not uniquely determined in the remaining table, and that it is determined by the two numbers b and c we add back, together with one number d which is in the same column. Since there is only one number in this column, we may as well suppose that c is not originally written; thus c can only be determined by two numbers e and f in another column. Therefore, a can be determined by e, f and d in the remaining table, which is a contradiction. Now we only need to find one column which satisfies the two above conditions, namely that it contains exactly one number. We notice that $m \leqslant n$, so the number of the numbers in the table is $r \leqslant k - 1 = m + n - 2 \leqslant 2n - 2 < 2n$. Thus there is at least one column which has not more than one number. If there is no number in this row, then this row is not uniquely determined, so the conclusion is true. If there is only one number a, then we delete this column, and notice that the remaining table can be uniquely determined. In fact, suppose b is not uniquely determined, then b will be determined by a, thus a and b are the two vertices of a rectangle, and that the three numbers besides b in this rectangle are all known, thus there must be one number e in the same column with a that is originally written, which is a contradiction. Therefore, the table of size $(m-1) \times n$ corresponds to the minimum of $\leqslant r - 1 \leqslant k - 2 = m + n - 3 < m + n - 2$, which is a contradiction.